KNOT THEORY

by

K. REIDEMEISTER

Originally published as KNOTENTHEORIE by K. REIDEMEISTER

Ergebnisse der Mathematik und ihrer Grenzgebiete,

(Alte Folge), Band 1, Heft 1.

Translated from the German and edited by LEO F. BORON, CHARLES O. CHRISTENSON, and BRYAN A. SMITH, all of the University of Idaho, Moscow, Idaho, USA.

BCS ASSOCIATES

Moscow, Idaho, U.S.A.

1983

Original German Edition Copyright © 1932 by Julius Springer in Berlin; Reprint of German Edition © 1974 by Springer-Verlag, Berlin-Heidelberg-New York

English Edition Copyright © 1983 by Leo F. Boron, Charles O. Christenson, and Bryan A. Smith; published by BCS Associates, Moscow, Idaho, USA.

Manufactured in the United States of America

All rights reserved.

No part of this book may be translated or reproduced in any form without written permission of the publisher.

ISBN: 0-914351-00-1

Library of Congress Catalog Card Number: 83-72870

CONTENTS

	Page
Foreword to the English edition	vi
Publisher's foreword to the original edition	vii
Introduction	x

CHAPTER I

Knots and their projections

§1.	Definition of a knot	1
§2.	Regular projections	3
§3.	The operations Ω. 1, 2, 3	6
§4.	The subdivision of the projection plane into regions	10
§5.	Normal knot projections	13
§6.	Braids	15
§7.	Knots and braids	19
§8.	Parallel knots. Cable knots	22

CHAPTER II

Knots and matrices

§1.	Elementary invariants	25
§2.	The matrices $(c_{\alpha\beta}^{h})$	27
§3.	The matrix (a_{ik})	31
§4.	The determinant of a knot	33
§5.	The invariance of the torsion numbers	35
§6.	The torsion numbers of particular knots	38

		Page
§7.	The quadratic form of a knot	40
§8.	Minkowski's units	46
§9.	Minkowski's units for particular knots	50
§10.	A determinant inequality	52
§11.	Classification of alternating knots	55
§12.	Almost alternating knots	57
§13.	Almost alternating circles	61
§14.	The L-polynomial of a knot	63
§15.	L-polynomials of particular knots	68

CHAPTER III

Knots and groups

§1.	Equivalence of braids	71
§2.	The braid group	73
§3.	Definition of the group of a knot	75
§4.	Invariance of the knot group	79
§5.	The group of the inverse knot and of the mirror image knot	83
§6.	The matrix $(\ell_{ik}(x))$ and the group	85
§7.	The knot group and the matrices $(c_{\alpha\beta}^{h})$	88
§8.	The edge path group of a knot	90
§9.	Structure of the edge path group	92
§10.	Covering spaces of the complementary space of the knot	96
§11.	The group of a parallel knot	99
§12.	The groups of torus knots	108
§13.	The L-polynomials of parallel knots	111
§14.	Several special knot groups	115

	Page
§15. A particular covering space	118
Table of knots	126
Bibliography	129
Index	135

FOREWORD TO THE ENGLISH EDITION

The reader might well ask why a translation is desirable 50 years after the original book was published. The answer is that Reidemeister's classic book has continued to be a useful introduction to a surprisingly large number of concepts.

We have (as much as possible) tried to retain the style of the original while trying to render the text into what we hope is readable English.

In a few places where modern terminology differs from Reidemeister we have changed to the current terminology. In a few places where errors in the original text were noted, we have added footnotes. An index has also been added.

We think that the reader will be amazed at the sheer wealth of material that is packed in this little book.

Our thanks go out to Springer-Verlag for licensing us to print this.

L.F.B.
C.O.C.
B.A.S.

Publisher's foreword to the original edition

The reporting of scientific literature is faced with a twofold problem: on the one hand, it must strive for completeness and timeliness with regard to the currently appearing journal literature, and on the other hand it must be concerned occasionally with the exposition of what has evolved over a period of time.

Whereas in all sciences it has been clear for a long time how in principle to master the first problem, fundamentally different attempts have been made to solve the second problem.

After almost thirty year's experience it is well established that the form of an encyclopedia which encompasses the collected field of mathematics and its impinging areas is not satisfactory. It is not likely that the undertaking begun here will do justice to all demands that one asks of a comprehensive reportage, but here nonetheless an attempt will be made to proceed with a fundamentally different method.

The series "Ergebnisse der Mathematik" will be as flexible as possible in order to follow the development of our science. It's goal is to introduce as individual independent monographs the problems, literature and the principal directions of development in particular modern areas. We desire also to take note of the actual state of things so that it will give each researcher the opportunity to acquire the reports which directly impinge on his area of interest, without forcing him to burden himself at the same time with all the particulars that are inherent in the somewhat excessive development of a comprehensive handbook of the entire science.

Accordingly, henceforth there will appear a series of reports covering all of the developing areas of mathematics and its closest applications. These will cover, initially, the approximately 6 or 7 shifts of direction in the research of the last decades. The frequency of later continuations will depend on the speed of progress. Taking a broad view, it is clear that no strongly formal unity of these reports can be aspired to; it will depend on the status of the available literature whether or not a purely literature report or a stronger textbook presentation is appropriate. The overall plan of the "Ergebnisse" series is to make available within a reasonable space of time, reports on almost all modern areas (at least of pure mathematics). Hence it will be possible to obtain the most comprehensive possible overview of the new developments of mathematics. In the combining of five reports into each volume, we will forego any objective grouping — after all, experience shows that all previous attempts at systematizing were a burden rather than a relief for the user.

The rules for citing literature which have recently been recognized internationally have been made use of.

The Editorial Staff of
Zentralbatt für Mathematik
und ihre Grenzgebiete

Introduction

Knot theory starts with the intuitive problem of deciding if two closed strings made of flexible, but impenetrable material can be transformed by means of continuous modifications into strings having congruent form. For example, if one ties a knot in an open string in the sense of ordinary language and then splices the two ends of the string, the result cannot be continuously deformed into a circle.

To formulate the given problem mathematically, mathematical representatives must be given for the strings. Then we must define deformations of the strings. This is naturally possible in various ways. For example, it is obvious to take for the strings double point free closed continuous curves in three-dimensional Euclidean space and to understand a deformation to be a continuous modification of this curve without self-piercings. More precisely: If x_1, x_2, x_3 are Cartesian coordinates and if

(1) $\quad \xi_i(t) = (x_{1i}(t), x_{2i}(t), x_{3i}(t)) \quad (i=1,2; 0 \leq t \leq 2\pi)$

are two disjoint curves given by parametric representation, which can be mapped by t in a 1-1 and continuous fashion onto the circle

(2) $\quad\quad\quad\quad\quad\quad x_1 = \cos t, \quad x_2 = \sin t.$

Then $\xi_1(t)$ is said to be deformable into $\xi_2(t)$ if there is a family of curves

(3) $$\xi(t,\tau) \qquad (0 \leq t \leq 2\pi, \ 0 \leq \tau \leq 1)$$

with

$$\xi(t,0) = \xi_1(t), \quad \xi(t,1) = \xi_2(t),$$

which are each mapped 1-1 and continuously onto the circle (2) by t and which sweep out a double point free surface $\xi(t,\tau)$. Instead of this embedding in families of curves, one can also use mappings of the entire space onto itself as a classification principle: The two curves (1) are called equivalent if there exists a 1-1 continuous mapping

$$x'_k = f_k(x_1, x_2, x_3) \qquad (k = 1,2,3)$$

of Euclidean space onto itself for which

$$x_{k2}(t) = f_k(x_{11}(t), x_{21}(t), x_{31}(t)) \qquad (k = 1,2,3).$$

But both formulations are too general since arbitrary continuous curves do not have intuitively obvious properties while special continuous curves, for instance, polygons suffice for the study of the intuitive curves. Accordingly, we will henceforth consider only polygons consisting of finitely many Euclidean line segments, and require that the family of curves (3) likewise consist only of polygons. The allowable transformation of the polygons can be composed of special deformations under which actually only one or two adjacent segments of the polygon are involved. Thus the conceptualization in §1, Ch. I, turns out to be a natural consequence of the intuitive point of view.

Tietze, in Monatsheft Math.-Phys., Bd 19, p. 34, gave an example that shows that there are continuous curves that cannot be deformed into polygons and that the classification problem formulated in (1) and (3) is more inclusive than that formulated in §1, Ch. I. Tietze's example is of a curve with infinite knotting. It arises from an infinitely often iterated tying together of polygon knots. The curves that are deformable into polygons can be embedded in three-dimensional tubes, which are made up of 3-spheres whose midpoints wander along the curves. They are thus in fact also suited as the representatives of intuitive strings, if of course they do not exceed a certain thickness.

One can investigate knot theory more deeply by seeking those properties of Euclidean space that are pertinent for the classification of knots. These are certain topological properties of the space: the knot problem is nothing other than the so-called isotopy problem for simple curves in the three-dimensional sphere. A treatment of knot theory as a part of isotopy theory, however, <u>in extenso</u> (at full length) has not yet been fully developed. The methods of combinatorial topology, for instance, are applied only in an attempt by Dehn to characterize the circle, and this work directly displays the particular difficulties with which one battles when using the combinatorial method. Thus the current definition of a knot with the aid of Euclidean polygons can be justified by the status of scientific development.

What is the present situation of knot theory? Elementary considerations starting with the most noticeable forms of the knot projections have not led to any proven results. It is easy to give necessary conditions for two curves to be topologically equivalent, but there is no success in applying these properties to a given curve. This difficulty first became apparent when Poincaré assigned certain groups to manifolds and thereby to the knots (the group of a knot is the fundamental group of the space which arises from the Euclidean space if the points of the knot are removed). Wirtinger and Dehn, to be sure, gave methods by which the generators and defining relations of this group can be read from the knot projection. That method led, for instance, to a proof that the trefoil knot is not a circle and that the trefoil knots cannot be deformed into their mirror image. But the answering of general questions was frustrated by the difficulty of evaluating the group-theoretic tools that Poincaré's discovery had placed in the hand of the geometer.

Thus it appears that the further development of knot theory is closely connected with the progress of group theory. In fact, one can obtain from the group of the knot most of the known knot properties by means of a purely group-theoretic algorithm. Namely, they can be obtained by the algorithm for determining generators and defining relations of subgroups. While one can obtain the torsion numbers and the Alexander L-polynomial of the knot without the direct utilization of this procedure, the derivation of their relationship to covering spaces leads in a completely natural way to the general considerations that yield a proof of the group-theoretic algorithm.

Besides the group there is the quadratic form assigned to the knot; from it comes the fact that the generators and defining relations of the knot group read from the knot projection are of a special nature not determined by the group structure.

It is not difficult to write a development of knot theory in which the group of the knot and the algorithm for determining the subgroups is placed in the forefront. For example, one can start directly with the geometrically meaningful definition of the knot group in Ch. III, §§8 and 9 and the determination of torsion numbers and L-polynomials in Ch. III,

§§ 6 and 7. This development even offers noticeable advantages since it reduces the invariance proof to the invariance proof for the group and to the problem of the invariant geometric interpretation of the torsion numbers and the L-polynomials. But I have preferred to work out the formal elementary character of the properties of a knot that are obtainable from matrices and to establish these interconnections for their own sake in Chapter II. This is in order to do better justice to the works of Alexander on the one hand and to the remarkable quadratic form of the knot on the other hand.

CHAPTER I

KNOTS AND THEIR PROJECTIONS

§1. Definition of a knot

In order to define the concept of a knot ([5],[28]),[*] we make use of simple closed polygons in Euclidean 3-space which consist of finitely many line segments. We understand a <u>deformation</u> of a polygon to be the generation of a new polygon from the original one by means of the following two operations:

Δ. Let $P_p P_1$ be an edge of the polygon with endpoints P_p and P_1, and let $P_p P_{p+1}$ and $P_{p+1} P_1$ be two line segments not belonging to the polygon, with the endpoints P_p, P_{p+1} and P_{p+1}, P_1 respectively. Assume that the surface of the triangle $P_p P_{p+1} P_1$ has no points in common with the polygon except for the edge $P_p P_1$. Then replace $P_p P_1$ by $P_p P_{p+1}$ and $P_{p+1} P_1$.

Δ' is the operation inverse to Δ: Suppose that the surface of the triangle $P_p P_{p+1} P_1$ formed from three consecutive vertices P_p, P_{p+1}, P_1 of the polygon has no point in common with the polygon except for the segments $P_p P_{p+1}$ and $P_{p+1} P_1$. Then $P_p P_{p+1}$ and $P_{p+1} P_1$ are replaced by $P_p P_1$.

---

[*]These numbers refer to the bibliography found at the end of this volume.

Polygons that arise from one another by a finite sequence of deformations are called <u>isotopic</u>, and a class of isotopic polygons is called a <u>knot</u>. Furthermore, we call a polygon itself a knot, although it would be more precise to call it a representative of a knot.

A property of a polygon that is preserved under the operation Δ or Δ' is called a <u>knot property</u> of the polygon. The task of knot theory is to obtain a survey of all properties of a knot, i.e., to find all deformation invariants of a simple closed polygon.

A knot that is isotopic to a triangle is called a <u>circle</u>. A polygon that is not isotopic to a triangle is said to be <u>knotted</u>.

A knot can be <u>oriented</u> by choosing a sense of direction for going around the knot. The classification problem can be extended to oriented knots. The knot that is oriented opposite to a given oriented knot is called the <u>inverse</u> knot; a knot is called <u>symmetric</u> if the two knots arising from it by orientation are isotopic. A knot is called <u>amphicheiral</u> if it is isotopic to its mirror image.

Instead of individual polygons, one can consider systems of finitely many closed, simple, and mutually disjoint polygons and apply the operations Δ and Δ' to the system. For these deformations it is reasonable to require that the surface of the triangle $P_p P_{p+1} P_1$ has no point in common with any of the polygons of the system except for the segments which are replaced under the deformation. Two such systems are called <u>isotopic</u>, and each represents the same <u>link</u>, if one can be transformed into the other by a finite sequence of the deformations Δ and Δ'. The simplest invariant of a link is the number of the polygons belonging to it; this number is called the <u>order</u> of the link.

§2. Regular projections

A regular projection of a polygon offers a convenient way to represent an individual knot.

By a <u>parallel projection</u> we understand the usual projection of Euclidean 3-space onto a 2-dimensional subspace. We say that the image of a polygon under a parallel projection is <u>regular</u> if each projecting ray meets at most two segments of the polygon. Hence the only singular points of the projection are double points, and no double point of the projection corresponds to a vertex of the polygon. One concludes from this that a regular projection can possess only finitely many double points. In order for a projection to have infinitely many double points, the projection of two segments of the polygon project into a segment s of double points, and the boundary of s which corresponds to vertices of the polygon, must also consist of double points. Accordingly, there are two cases for how a projection direction can be singular:

a) There are no double points that correspond to vertices of the polygon. In this case, there are higher order singular points. Then the projection ray through this higher order singular point meets at least three lines on which segments of the polygon lie. These lines must be skew since otherwise a singular point would occur that corresponds to a vertex of the polygon; and the projection ray therefore belongs to the one-sheeted hyperboloid determined by the three lines. For each triple of skew lines on which segments of the polygon lie, we consider the hyperboloid determined by them. Then we form the quadric cones with vertex at the origin whose generating lines are parallel to the generating lines of these hyperboloids. All singular directions of this type are contained among the directions obtained.

b) The projection of a vertex falls upon the projection of a segment or upon the projection of another vertex. In this case the projection ray is parallel to a plane which passes through a segment and a vertex of the polygon which is not an endpoint of that segment.

The regular projection directions thus decompose into finitely many regions whose boundaries lie on the cones and the planes of the singular projection directions. A closer investigation of the singular projection directions can be of interest. This is shown, for example, by the following theorem about chords which meet a polygon in four distinct points ([26]). If c is the knotting number of the polygon (cf. §1, Ch. II), and if c is even, then the number of chords that meet the polygon in four distinct points is at least c^2. A similar theorem holds for a link consisting of two polygons k_1 and k_2. If c_{12} is the linking number of k_1 relative to k_2 and if c_{21} is the linking number of k_2 relative to k_1 (cf. §1, Ch. II), then the number of chords that meet the link in four distinct points is at least $c_{12}c_{21}$.

A regularly projected curve is decomposed by its double points $D_1, D_2, \ldots, D_n$ into $2n$ singularity free edge paths $z_1, z_2, \ldots, z_{2n}$, and these arcs decompose the projection plane into finitely many polygons $\Gamma_1, \Gamma_2, \ldots, \Gamma_g$ and one unbounded region Γ_0. By the Euler-Poincaré formula, $g = n + 1$. The boundary relations between D_i, z_k, Γ_ℓ, i.e., the designation of which two double points bound z_k and of which two regions have z_k in their boundary, we call the <u>schema of the projection</u>.

In order to obtain the knot from the projection, it is necessary to know which of the arcs at the double points correspond to overcrossings and which correspond to undercrossings. We fix for each projection direction a sense of above and a sense of below and designate the points on the knot corresponding to the double point D_i of the projection by U_i and U^i. In this, U_i lies under U^i and U_i is called an <u>undercrossing point</u>, while U^i is called an <u>overcrossing point</u>. We <u>normalize</u> the projection by specifying for each D_i which of the edge paths z_k emanating from D_i are projections of edge paths emanating from U_i and U^i respectively (Figs. 1 and 2). We can thus normalize the schema of the projection.

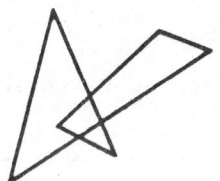

Fig. 1.   Fig. 2.

If two polygons have the same normalized projection, then they are isotopic. If two polygons k_1 and k_2 both project to the same curve, but if the normalization of their projections is reversed at each of the double points, then k_1 is isotopic to the mirror image of k_2.

We say that a knot projection is <u>alternating</u> if each edge path joins an overcrossing to an undercrossing at an adjacent double point, therefore upon transversing the knot the overcrossing and undercrossing points alternate (Fig. 2). We also speak of alternating portions of a projection. A regular projection can always be normalized to be alternating, and indeed in precisely two ways. The corresponding knots are then mirror images (Fig. 2). Accordingly, in the table of knots with up to nine double points given on pages 126-128, the unnormalized projections always signify alternating projections. Knots for which there exist alternating projections are called <u>alternating</u> knots.

§3. The operations $\Omega.1,2,3$

We will now investigate how the normalized projection is altered by deformations of the polygon and by changing the direction of the projection ([5],[28]).

We first list several types of modifications which are effected by knot deformations.

$\Delta.\pi.1$. This is the alteration of the projected curve by the operation Δ or Δ'. The projected curve is of course also a polygon--which might even have double points--to which these transformations can be applied.

$\Delta.\pi.2$. This is the application of Δ or Δ' in the following setting. Let the triangle $P_p P_{p+1} P_1$ appearing in the deformations Δ or Δ' project to a triangle $P'_p P'_{p+1} P'_1$ whose boundary is met by the remaining segments of the projected curve in precisely two points D and D' on the segments $P'_p P'_1$ and $P'_p P'_{p+1}$ respectively and which contains no double points in its interior (Fig. 3).

$\Delta.\pi.1,2$ are called <u>deformations of the projected curve</u>. They do not change the schema of the projection. Furthermore, one observes that projected curves with the same schema can be carried into one another by means of a sequence of deformations $\Delta.\pi.1,2$. From this it

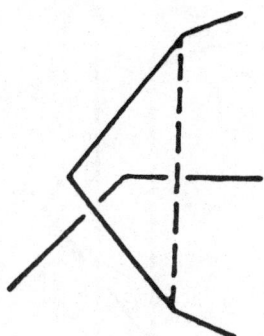

Fig. 3. Δ.π.2.

follows that polygons whose projections possess the same normalized schema are isotopic.

Furthermore, the following three operations which change the schema of the projection may arise from the operations Δ.

Ω.1. An edge path whose projection was double point free is transformed into a loop. In this a new double point is introduced. The corresponding undercrossing point and the overcrossing point on the polygon are adjacent (Fig. 4).

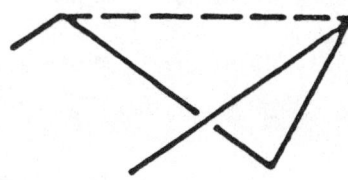

Fig. 4. Ω.1.

Ω.2. If we have two edge paths of the knot whose projections have no points in common, then one edge path is slid over the other so that there appear two adjacent overcrossing points in one edge path and two adjacent undercrossing points in the other edge path (Fig. 5).

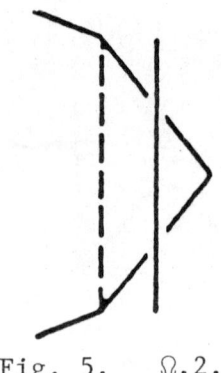

Fig. 5. $\Omega.2.$

$\Omega.3.$ Initial configuration: Three edge paths of the knot s_1, s_2, s_3 which yield three double points in the projection. Each double point is adjacent to the other two. Assume s_1 crosses over s_2 as well as s_3, and s_2 crosses over s_3. Operation: s_1 is pushed over the undercrossing point and the overcrossing point determined by s_2 and s_3 (Fig. 6).

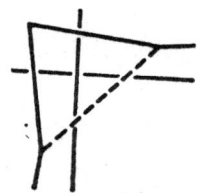

Fig. 6. $\Omega.3.$

We denote the inverse operations by $\Omega'.i$ $(i = 1, 2, 3)$. It can now be shown that each change of the projection which is induced by the knot deformations Δ and Δ' can also be generated by repeated application of the deformations $\Delta.\pi.1,2$ and the operations $\Omega.i$ $(i = 1,2,3)$ together with their inverses.

We sketch a proof of that assertion. Suppose that the knot is deformed by replacing the segment $P_p P_1$ by the two segments $P_p P_{p+1}$

and $P_{p+1}P_1$. Assume that the projections P'_p, P'_{p+1}, P'_1 of P_p, P_{p+1}, P_1 do not lie on a line. The projection direction is chosen so that the original as well as the deformed knot are projected regularly. The triangle $P'_p P'_{p+1} P'_1$ clearly contains only finitely many double points in its interior and on its boundary. The triangle can therefore be subdivided by segments which are parallel to $P_p P_1$ and $P_p P_{p+1}$ into triangles and parallelograms, so that the corresponding triangles and parallelograms of the projection each contains at most one double point in its interior. If a triangle or parallelogram contains a double point in its interior, then it is intersected by exactly four edge paths z_i, otherwise it is intersected by at most one edge path z_i ([1],[2]).

Now, by means of finitely many applications of the operations $\Delta.\pi.i$, $\Omega.i$, and $\Omega'.i$ one can replace $P_p P_1$ by $P_p P_{p+1}, P_{p+1} P_1$ and conversely, by "step by step" use of the triangles and quadrangles of the subdivision.

The changes of the projection induced by shifting the direction of the projection can also be given in terms of the operations $\Delta.\pi.$ and $\Omega.$ If the projection direction is varied continuously so that it always remains regular, then the changes of the projection can be generated by the $\Delta.\pi.$ In order to pass over a particular singular projection direction, we deform the knot so that this direction is no longer singular, we then pass over the direction, and then deform the knot back to its original form.

Thus we have shown: The knot properties coincide with those properties of the normalized schema which are preserved under the operations $\Omega.i$, $\Omega'.i$ ($i = 1,2,3$). Therefore, the knot problem is equivalent to determining when a normalized regular projection cannot be changed into another by a finite sequence of applications of the operations Ω. The analogous situation holds for links.

§4. The subdivision of the projection plane into regions

The regions Γ of the projection plane can be decomposed into two classes, for instance, by coloring them black or white, so that regions that are adjacent along an arc have different colors. Then regions lying on opposite sides of a double point have the same color.

This decomposition does exist. For, if a simple closed polygonal curve in the projection plane intersects the knot's projection finitely often, say c times, and if it does not pass through a double point, then c must be an even integer.

Let the non-compact region always be colored black. The collection of white regions can then be conceived of as the projection of a compact surface, whose boundary is the knot. The surface obtained this way can be either orientable or non-orientable. However, it can be shown that every knot is the boundary of some orientable surface ([18]).

Fig. 7.

Consistent with later usage, we will call a subset of a knot projection a <u>braid with two strings</u> if the subset bounds a sequence of black or a sequence of white regions, each of which has at most two double points in its boundary. Furthermore, we require that at each double point two of these regions are adjacent (Fig. 7). Using $\Omega'.2$ one can show that a braid with two strings can either be eliminated or be made alternating.

The simplest knots in the sense of black-white coloring are those "alternating torus knots" whose projections have exactly two black regions. They can be placed on a polyhedron that is topologically equivalent to the torus, and they bound on the torus a multiply twisted band that is topologically equivalent to a Möbius band (Fig. 7). An alternating torus knot with three double points is called a <u>trefoil knot</u> (or a "<u>cloverleaf knot</u>"). Alternating torus links of two polygons are defined analogously.

We define as <u>pretzel knots</u> that class of knots whose projections contain exactly three black regions. The overcrossings can in that case be divided into three two-stringed braids (Fig. 8).

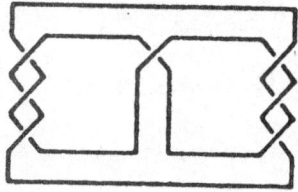 

Fig. 8.   Fig. 9.

Suppose that there is a white region Γ_1 that borders a black region Γ_2 along two distinct edge paths which are separated from one another by double points (Fig. 9). In this case let $w = w_1 w_2$ be a simple polygonal closed curve which intersects one of these segments once at P and the other once at Q. Furthermore, assume that w_i is situated entirely in Γ_i (i = 1,2). Let k_1' be the part of the projection lying in the exterior of w, and let k_2' be the part lying in the interior of w. If we now adjoin to each of the k_i' the path w_1, then there arise two knots k_i (i = 1,2) which are designated as <u>composite parts</u> of the original knot, if k_1 as well as k_2 are knotted. It is not difficult to reduce the properties of a knot to those of its component parts ([3],[14]). In general, however, a knot does not possess composite parts.[*] In the table of knots having at most nine overcrossings given on pp. 126-128, the knots with two or more composite parts are omitted.

If a double point D is incident twice with the same region Γ then the two corners belonging to Γ must lie crosswise at D since they are colored the same. Such a double point can be eliminated by a deformation of the knot (Fig. 10). Namely, there is a closed path w which is situated entirely in Γ and which intersects the knot projection exactly once, at D. If one rotates the part of the knot projection bounded by this path in a suitable way through 180°, then one obtains a projection which no longer contains the double point D. This

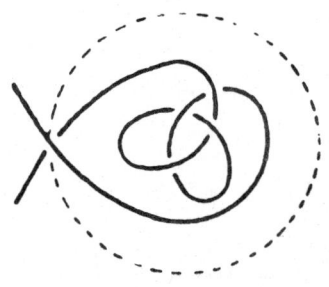

Fig. 10. $\Omega.4$.

---

[*] A knot with no composite parts is called <u>prime</u>.

operation--call it $\Omega.4.$--can be induced by a sequence of knot deformations, as one easily sees from considering the corresponding transformations of the knot in 3-space.

By means of repeated application of $\Omega.4.$ one may assume that the knot projection has no double points which are incident twice with the same region. Note that: An alternating projection remains alternating under $\Omega.4.$ To wit, in the part of the projection enclosed by the path w the overcrossings are turned into undercrossings and conversely; at D one overcrossing point and one undercrossing point are eliminated.

§5. Normal knot projections

A knot projection is called <u>normal with respect to the two black (white) regions</u> Γ_i, Γ_k if any of the following hold:

1) Γ_i and Γ_k do not have a common double point,
2) Γ_i and Γ_k are incident at exactly one common double point, or
3) all double points at which Γ_i and Γ_k are both incident lie on one alternating two-stringed braid.

The projection is called <u>normal</u> if it is normal with respect to each pair of regions that are colored the same, and no region is adjacent to itself.

Each regular normalized knot projection with n double points can be transformed into a normal projection with at most the same number of double points.[1]

---

[1] L. Goeritz, <u>Bemerkungen zur Knotentheorie</u>, Abh. Math. Sem. Hamburg 10 (1934) 201-210.

Suppose that a projection is not normal and, for example, that two double points D_1 and D_2 are incident to different two-stringed braids of the projection, both of which lie between the two black regions Γ_i and Γ_k. Then one can obtain a path $w_i w_k$ in the projection plane which intersects the projection only at D_1 and at D_2. Assume that w_i runs from D_1 to D_2 in Γ_i and that w_k runs from D_2 to D_1 in Γ_k. Now by means of $\Delta.\pi.$ the projection as well as $w_i w_k$ may be deformed so that $w_i w_k$ becomes a circle which is bisected by D_1 and D_2. Then rotate the part of the knot lying inside the circle $w_i w_k$ through 180° using the line through $D_1 D_2$ as axis (Fig. 11). In this the direction of rotation is arranged so that at D_2 an overcrossing is removed and at D_1 an undercrossing is added. Call this operation $\Omega.5$.

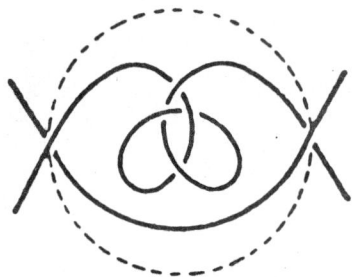

Fig. 11. $\Omega.5$.

One can see from the corresponding isotopy in 3-space that $\Omega.5$ can be induced by a sequence of knot deformations. The choice of which two-stringed braid will contain the two overcrossings after application of $\Omega.5$ is arbitrary and will correspond to a selection of the direction of rotation.

We see that by a repeated application of $\Omega.5$, every projection can be transformed into a normal projection relative to the black regions as follows: The double points of the part of the projection enclosed by the path that are incident with Γ_i and Γ_ℓ go over into double points that are incident with Γ_k and Γ_ℓ and those incident with Γ_k and Γ_ℓ go over into those incident with Γ_i and Γ_ℓ. Therefore, if the projection is normal with respect to Γ_i and Γ_ℓ, then the changed projection is normal with respect to Γ_k and Γ_ℓ. All other incidences relative to the black regions are retained. In particular, all double points that are incident with Γ_i and Γ_k are changed into double points that are incident with Γ_i and Γ_k. Therefore one can first make the knot projection normal with respect to Γ_i and Γ_k and then repeat the construction for all remaining pairs of black regions. Then $\Omega'.2$ can be used to change each braid into an alternating one. In the above, two normal white regions with two or more common double points remain normal, and there arise no new non-normal white region pairs. Therefore to obtain a normal projection, one applies the above process successively for all pairs of black regions and for all pairs of white regions.

The proof that alternating knots are transformed into alternating knots by $\Omega.5$ is analogous to the proof of this assertion for $\Omega.4$.

§6. Braids

We understand an open braid ([7]) with q strings to be the following structure: Let two congruent equidistant sequences of points $A_1, A_2, \ldots, A_q$ and $B_1, B_2, \ldots, B_q$ respectively be marked on a pair of opposite sides g_1 and g_2 of a rectangle P. Assign to each of the points A_i a unique point B_{k_i} and connect A_i with B_{k_i} by a simple edge path directed from A_i to B_{k_i}; each pair of these edge paths are disjoint. The projections of the edge paths on the plane determined by g_1 and g_2 should furthermore be situated entirely in the interior of the rectangle formed by g_1 and g_2 and be met in at most one point by each line parallel to g_i. Further, suppose at most one double point of the braid projection (Fig. 12) lies on each line parallel to g_i.

We call the q edge paths the strings of the braid. If g is a line parallel to g_i, if g goes through the double point D of the projection, and if g meets say k-1 strings to the left of D (hence there are exactly q-k-1 strings of the projection to the right of D), then we call D a crossing of the k-th and (k+1)-st strings. The numbering of the strings changes at each double point. We say that a braid is <u>twisted uniformly</u> (<u>gleichsinnig verdrillt</u>) if the k-th string always crosses over (or under) the (k+1)-st string.

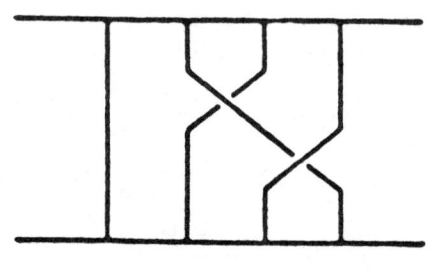

Fig. 12.

Cylindrical braids are examples of braids that are twisted uniformly. By a <u>cylindrical braid</u> with twisting +1 (-1) we understand a braid with q-1 overcrossings in which consecutively the i-th string crosses over (crosses under) the (i+1)-st string (i = 1,2,..., q-1). If a braid can be decomposed by $|r|$ - 1 parallels to the rectangle sides g_i into $|r|$ cylindrical braids each with twisting

+1 (−1), then the braid is called a cylindrical braid with twisting r > 0 (r < 0). Such braids can clearly be embedded on the surface of a cylindrical polyhedron.

The two-stringed braid introduced in §4, Ch. I, can be changed by the operations Δ.π. into an open two-stringed cylindrical braid.

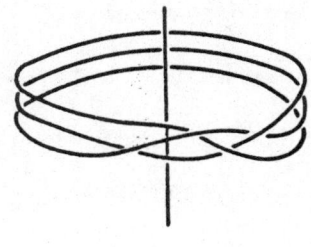

Fig. 13.

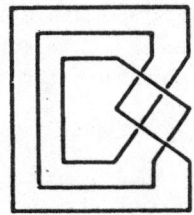

Fig. 14.

By a <u>closed braid</u> with the line <u>a</u> as axis we understand a simple closed, oriented polygon, or a system of such polygons, whose orthogonal projection is regular in a plane that is perpendicular to <u>a</u>, and which has the following further properties: If P and Q are two consecutive polygon vertices in the sense of the orientation, and if A is the point of intersection of the axis <u>a</u> and the projection plane, then APQ must be the positive orientation of the triangle APQ. The polygons wind around the axis in a definite direction <u>in toto</u> a finite number, say q times; q is called the <u>order</u> of the braid (Fig. 13).

A closed braid can be assigned to an open braid in the following way (Fig. 14): We choose the plane containing the rectangle P as the projection plane and choose as axis a line <u>a</u> which is perpendicular to the projection plane and pierces it in a point A that is outside the rectangle P. We then join each A_i and B_i with a simple edge path in the projection plane that is situated as follows: each of these edge paths is outside of P, each pair of edge paths are disjoint, and if the line segment $A_i B_i$ is added to the i-th edge path one obtains a convex polygon that contains A in its interior. One then erases the sides of the rectangle. An open two-stringed braid that is twisted uniformly thus becomes an alternating torus knot or an alternating torus link. In general, a closed braid that corresponds to a cylindrical braid can be embedded on a polyhedron that is topologically equivalent to the torus (Fig. 14), and is accordingly called a <u>torus braid</u>, a <u>torus knot</u>, or a <u>torus link</u>.

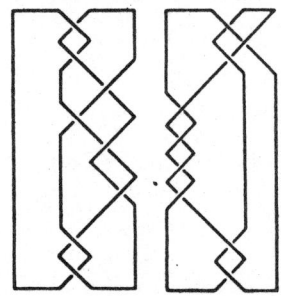
Fig. 15.

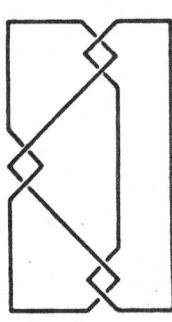
Fig. 16.

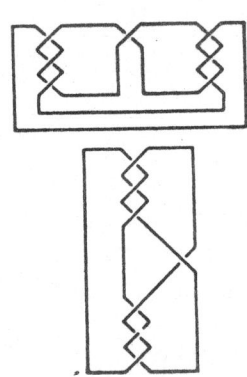
Fig. 17.

By operations $\Delta.\pi.$ one sees that any closed braid is equivalent to a closed braid that is assigned to some open braid.

One may also obtain knots or links from an open braid in other ways. For example, suppose a four-stringed open braid is given. We connect the two first and two last points on the top and respectively on the bottom of the rectangle P and erase the remaining perimeter of the rectangle. The system arising in this way which consists of one or two polygons is called <u>a plait with four strings (Viergeflecht)</u>.

Figure 15 depicts the knots 8_{14} and 9_8 of the knot table on pp. 126-128 as plaits with four strings; Fig. 16 is a link that is represented as a plait with four strings.

We can always deform a plait with four strings so that the uppermost a_1 crossings lie on the two middle strings, followed by $b_{1\lambda}$ crossings on the two left strings and then $b_{1\rho}$ crossings on the two right strings, this is followed by a_2 crossings in the middle, $b_{2\lambda}$ and $b_{2\rho}$ crossings on the left and right, respectively, and so forth, ending finally with a_n crossings on the two middle strings. Therefore, the double points of a plait with four strings can be subdivided into $3n-2$ two-stringed braids, which by §4, Ch. I, we may assume are alternating. Thus the schema of a plait with four strings is characterized by the numbers a_i, $b_{i\lambda}$, $b_{i\rho}$ together with specification of the sense of twisting for each two-stringed braid part. An analogous assertion holds for plaits consisting of $2a$ strings. Those pretzel knots that have only one crossing on the middle two-stringed braid part can easily be deformed into a plait with four strings (Figs. 8 and 17).

§7. Knots and braids

An arbitrary knot can be deformed into a braid ([2]). To prove this, consider any regular knot projection with vertices P_i ($i = 1, 2, \ldots, p$). The knot projection is formed by the line segments $s_i = P_i P_{i+1}$ ($i = 1, 2, \ldots, p$), where $P_{p+1} = P_1$. Let A be any point. We will display the projection in a braid-like manner about A. Assume that none of the triangles

$$AP_i P_{i+1} \quad (i = 1, 2, \ldots, p)$$

is degenerate. We split the segments into two classes, "positive" and "negative," depending on whether AP_iP_{i+1} is the positive or negative orientation for the triangle AP_iP_{i+1} (as a subset of the oriented plane). We claim that we can use an operation to remove a "negative" segment s_i without introducing new segments of this type.

This is clear when s_i contains at most one double point. Namely, if AP_iP_{i+1} is the negatively oriented triangle corresponding to s_i and if $A'P_iP_{i+1}$ is a slightly altered triangle that contains A in its interior, then we can replace the segment s_i by the edge path $P_iA'P_{i+1}$. In doing this we normalize the double points lying on $P_iP'P_{i+1}$ so that $P_iA'P_{i+1}$ crosses over (under) all segments that it meets, depending on whether s_i contains an overcrossing point (undercrossing point). If s_i contains no crossing points then either possibility will do. Now, if s_i contains k double points $(k > 1)$, then we introduce new vertices $P_{i1}, P_{i2}, \ldots, P_{i,k-1}$ in order to decompose s_i into the segments $s_{i1}, s_{i2}, \ldots, s_{ik}$ each of which contains only one double point. Now we proceed with $s_{i\ell}$ as we did previously with s_i. Since no new double points are introduced on $s_{i\ell}$ when we eliminate s_{ij} ($j \neq \ell$) we can eliminate s_i in k steps. An analogous theorem holds for links.

That one knot can be related to different braids is shown for example by Figs. 14 and 7 which both represent the trefoil knot. We emphasize this by stating the following theorem[1]: If a braid that is twisted uniformly has exactly two overcrossings of the k-th and

---

[1] C. Bankwitz, Über Knoten und Zöpfe in gleichsinniger Verdrillung, Math. Zeit. 40 (1935) 588-591.

$(k + 1)$-st strings and fewer than four overcrossings of the $(k + 1)$-st and $(k + 2)$-nd strings, then it can be deformed into a braid of fewer strings which is twisted uniformly.

We make two additional observations about plaits with four strings.[1] Every plait with four strings v can be deformed into normal plaits with four strings v', respectively v'', which have the following property:

$v'(v'')$ possesses only overcrossings on the two middle strings and the two left (right) strings. If $a_i, b_{i\lambda}, b_{i\rho}; a'_i, b'_{i\lambda}, b'_{i\rho}; a''_i, b''_{i\lambda}, b''_{i\rho}$ $(i = 1, 2, \ldots, n)$ are the numbers corresponding to the plaits v, v', and v'', then we have $b'_{i\rho} = 0$, and $b''_{i\lambda} = 0$.

Furthermore,

$$a_i = a'_i = a''_i \quad \text{and} \quad b'_{i\lambda} = b''_{i\rho} \quad (i = 1, 2, \ldots, n).$$

If v is alternating, then v' and v'' are also alternating, and the corresponding two-stringed braid parts are all twisted uniformly. In addition, we have that

$$b'_{i\lambda} = b_{i\lambda} + b_{i\rho} = b''_{i\rho} \quad (i = 1, 2, \ldots, n).$$

If v is not alternating, then the relations between the twisting of the corresponding two-string braid parts and the numbers are somewhat more complicated.

---

[1] C. Bankwitz, <u>Über Knoten und Zöpfe in gleichsinniger Verdrillung</u>, Math. Zeit. <u>40</u> (1935) 588-591.

v' and v" arise from one another by a rotation through 180° and possibly deformations Δ.π. From this it can be deduced that the knots that are representable by a plait with four strings v are symmetric.

§8. Parallel knots. Cable knots

Let x_i (i = 1,2,3) be cartesian coordinates, z a closed braid with the x_3-axis of the coordinate system as the braid axis, and let $x_3 = 0$ be the equation of the projection plane; then for d sufficiently close to one, the transformation

$$x_1' = dx_1, \quad x_2' = dx_2, \quad x_3' = x_3,$$

sends z to an isotopic braid $z^{(1)}$, which has no point in common with z. One sees that the projection of $z^{(1)}$ is an outer or inner parallel of the projection of z, according as $d > 1$ or $d < 1$. $z^{(1)}$ itself is a parallel of z. One can iterate this process say q times, thus obtaining q braids $z^{(1)}, z^{(2)}, \ldots, z^{(q)}$, which collectively form a link consisting of q isotopic curves.

In order to visualize this link in another way, we consider a tube S which is the envelope of all balls with constant radius ρ whose midpoints lie on z. If ρ is small enough, then S is a singularity free surface, which can be mapped homeomorphically onto a torus. We shall call z the _core_ of S. We can now deform the $z^{(i)}$ (i = 1,2,...,q) into a system of simple curvilinear curves on the tube S.

We now apply the following transformation to the link so obtained. Let s_i be an edge of z and let $s_i^{(k)}$ ($k = 1, 2, \ldots, q$) be the corresponding parallel edges of $z^{(i)}$. Let these edges intersect the opposite sides g_1, g_2 of a rectangle in $A^{(k)}$, $B^{(k)}$ and let the q segments $A^{(k)}B^{(k)}$ be replaced by a cylindrical braid of order q and twisting r. If q and r are relatively prime, then there arises a single polygon which is called a knot parallel to z (or, briefly, a _parallel_ knot) (Fig. 18). If one forms a parallel knot by cutting at a different place and adjoining a cylindrical braid with twisting r, then there arises a parallel knot that is isotopic to the first one. Hence, the isotopy class of a knot parallel to z is determined by the numbers q and r.

Fig. 18.

Note that we have not claimed that distinct pairs of numbers yield distinct isotopy classes.

Similarly, one can define parallel knots k_{qr} for an arbitrary knot k. Moreover, let it be noted that: If one deforms k into k' and forms the parallel knots, $k'_{qr'}$, of k', then each knot k_{qr} can be deformed into at least one knot $k'_{qr'}$; in this, under certain conditions, parallel knots with different r, r' may correspond to one another. All the simple closed curves on the tube S can be deformed into knots that are parallel to the core of the tube, or into a triangle.

We define <u>cable knots</u> ([11],[21],[34]) of the s-th degree to be those knots which arise by the s-fold formation of parallel knots from the circle.

A cable knot of the first degree is therefore determined by giving two relatively prime numbers q_1 and r_1. The q_1 strings wind around the circle $|r_1|$ times. These knots are called torus knots since they can be drawn as a simple closed curve on the surface of an unknotted torus.

A cable knot of the second degree lies on the tube whose core is a torus knot; it is determined by giving the four numbers q_1, r_1, q_2, r_2:

q_1, r_1 to determine the torus knot,

q_2, r_2 to determine their parallel knots

(Fig. 18).

In general, a cable knot of the s-th order is determined by giving s number pairs

$$q_1, r_1; q_2, r_2; \ldots; q_s, r_s,$$

wherein $q_i > 1$ and q_i, r_i must be relatively prime. The cable knots have an important and close relationship with the singularities of plane algebraic curves ([11],[21],[34]).

CHAPTER II

KNOTS AND MATRICES

§1. Elementary invariants

It is very easy to define a number of knot invariants so long as one is not concerned with giving algorithms for their computation. For instance, among all regular projections of a knot, there are those in which the number of double points, regions, or black regions of the projection is the smallest and, hence, by their definition, the corresponding minimal numbers are knot variants.

One can change each knot projection into the projection of a circle by reversing the overcrossings and undercrossings at, say, k double points of the projection. The minimum number $m(k)$ of these operations, that is, the minimal number of self-piercings, by which a knot is transformed into a circle, is a natural measure of knottedness.* Furthermore, by §4, Ch. I, we can span any polygon by at least one double point free polyhedron. To each double point free polyhedron we can assign the Euler-Poincaré characteristic k; the minimum $m(k)$ of the Euler-Poincaré characteristics over all surfaces that span the knot is an invariant. A circle is clearly characterized by the fact that spanning polyhedra of minimal characteristic are disks.

Finally, for each polygon one can consider the polyhedra with self-intersection that are bounded by the polygon. In combinatorial topology these are disks with singularities. Let r be the number of

---

*Wendt, H., <u>Die gordische Auflösung von Knoten</u>, Math. Z., 42 (1937), 680-696.

points that the polygon has in common with a spanning singular disk; we also call r the number of boundary singularities. Now, let the knotting number c be the minimum $m(r) = c$. The assertion that the circle is characterized by $m(r) = 0$ is the content of the so-called Dehn Lemma, whose proof is not yet valid ([15],[22]).* It is conjectured that c is always an even number; it follows from $c < 2$ that $c = 0$ ([26]).

However simple and geometrically significant the definition of these various knot properties may be, there is no method for computing them in general. For instance, there is no way to obtain them from a knot projection.

Some similar questions concerning links are somewhat more accessible. If a regular projection is given of a link of two oriented polygons k_1 and k_2, let D_i ($i = 1,2,\ldots,h$) be the double points in which k_1 crosses over the polygon k_2, and let ε_i be the characteristics assigned to these double points in (1), §2, Ch. II. If we set ([12])

$$\sum_{i=1}^{n} \varepsilon_i = v_{12}$$

and define v_{21} analogously, then $v_{12} = v_{21} = v$. The number v is called the <u>intertwining number</u>** of the link. The invariance of v under deformations is easy to verify; furthermore, v is invariant with respect to deformations by which the polygons k_i pass through themselves, but

---

*Proven by C. D. Papakyriakopoulos, <u>On Dehn's lemma and the asphericity of knots</u>, Ann. of Math. <u>66</u> (1957), 1-26.

**Rolfsen calls this the <u>linking number</u>. See D. Rolfsen, <u>Knots and Links</u>, Publish or Perish, Inc., Berkeley, 1976.

not each other. Using such deformations one can transform a link with the intertwining number v into an alternating torus link with $2v$ overcrossings. Gauss ([17]) gave an integral for calculating v; in order to calculate v by means of the edge path group of the knot k_1 or of the knot k_2, see §9, Ch. III.

Two polygons k_1, k_2 of a link are called <u>unlinked</u> if they can be transformed by deformations Δ and Δ' into polygons whose projections are disjoint.

If we apply to k_1 an arbitrary deformation in which we allow k_1 to intersect both itself and k_2, then k_1 can be transformed into a curve that is unlinked with k_2. By the <u>linking number</u> c_{12} of k_1 relative to k_2 we understand the minimal number of intersections with k_2 that are required to transform k_1 into a curve that is not linked with k_2.

In general, c_{12} is different from c_{21} ([26]); on the other hand, for unknotted curves we have $c_{12} = c_{21}$. We will show in §9, Ch. III, that there exist polygons with $v = 0$ and $c_{12} \neq 0$. There are plaits with four strings ([6]) which show that two polygons with $c_{12} = c_{21} = 0$ may still be linked; it is shown in §14, Ch. III, that the edge path groups of such links are different from the edge path groups of two linked circles. There is no known algorithm for calculating the linking number directly.

§2. The matrices $(c_{\alpha\beta}^h)$

We will now seek methods for obtaining knot invariants that can be computed. We proceed by giving first the purely formal calculation rules for the definition of the invariants, second we formally prove their invariance, and finally we give their geometric interpretation in §8, Ch. III, and §10, Ch. III.

Initially, we define matrices which can be read off directly from a given normal regular knot projection. The elementary divisors of these matrices that are different from one will turn out to be knot invariants.

Suppose the knot projection has n double points D_i ($i = 1, 2, \ldots, n$). The corresponding undercrossing points U_i decompose the knot into n edge paths $s_1, s_2, \ldots, s_n$. After designating a positive direction for traversing the knot, we can number the D_i and s_i so that s_i goes from D_{i-1} to D_i. The line segment $s_i s_{i+1}$ will be crossed over by $s_{\lambda(i)}$ at D_i ($i = 1, 2, \ldots, n$). In this, $D_0 = D_n$, $D_{n+1} = D_1$, and $s_{n+1} = s_1$.

After designating an orientation in the projection plane, we assign to the point D_i the <u>characteristic</u>

(1) $$\varepsilon = \pm 1,$$

where $\varepsilon = +1$ or $\varepsilon = -1$ depending on whether or not the directed overcrossing arc can be carried into the direction of s_{i+1} (Fig. 19) by a positive rotation about D_i through an angle smaller than a straight angle (Fig. 19).* The characteristic does not depend on the orientation of the knot.

---

*Note that throughout one assumes that a clockwise rotation in the plane is positive.

Ch. II, §2   THE MATRICES $(c_{\alpha\beta}^h)$

Now for each integer $h \geq 1$ we form a matrix $(c_{\alpha\beta}^h)$ with $h \cdot n$ rows and columns by use of the following rules ([8]):

To each arc s_i $(i = 1,2,\ldots,n)$ we assign h columns (i,k) $(k = 0,1,\ldots,h-1)$ and to each point D_i $(i = 1,2,\ldots,n)$ we assign h rows (i,k) $(k = 0,1,\ldots,h-1)$ of our matrix.

If $\lambda(i) \neq i, i+1$ and $\varepsilon_i = +1$, then we set in the row (i,k),

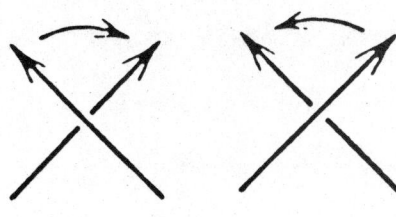

-1 in the column $(i + 1, k - 1)$*

$+1$ in the column (i,k)

$+1$ in the column $(\lambda(i), k-1)$

-1 in the column $(\lambda(i), k)$.

Fig. 19.

If $\lambda(i) \neq i, i + 1$ and $\varepsilon_i = -1$, then we set in the row (i,k),

-1 in the column $(i+1, k-1)$ | -1 in the column $(\lambda(i), k - 2)$

$+1$ in the column $(i, k - 2)$ | $+1$ in the column $(\lambda(i), k - 1)$

and in both cases a zero in the other places of this row.

If $\lambda(i) = i$, then for both $\varepsilon_i = +1$ and $\varepsilon_i = -1$, we place

-1 in the column $(i + 1, k - 1)$

$+1$ in the column $(i, k - 1)$

and zero in the remaining places.

If $\lambda(i) = i + 1$, then we write for $\varepsilon_i = +1$, respectively $\varepsilon_i = -1$,

$+1$ in column (i,k), respectively $(i, k - 2)$

-1 in column $(i + 1, k)$, respectively $(i + 1, k - 2)$

and zero in the remaining places.

---

*(i,-1) and (i,-2) signify the column $(i, h-1)$ and $(i, h-2)$ respectively if $h \geq 2$ and the column $(i,0)$ if $h = 1$.

The matrices $(c_{\alpha\beta}^h)$ can be read off from the normalized bounding relations (i.e., the schema of the projection) between the double points D_i and the edge paths s_k of the projection. But also, conversely, this normalized schema of the projection can be determined from each matrix $(c_{\alpha\beta}^h)$. The elementary divisors of $(c_{\alpha\beta}^h)$ that are different from one are called the <u>h-th torsion numbers</u>.

One obtains from the matrix $(c_{\alpha\beta}^h)$ a matrix with the same elementary divisors if for arbitrary i_0 one removes the h columns $(i_0,k)(k = 0,1,\ldots,h-1)$; this holds since if one adds together the n columns (i,k) with fixed k, one obtains a column consisting of zeros only.

The elementary divisors of the matrix $(c_{\alpha\beta}^1)$ are all equal to 1.

From the same incidence relations one can obtain another matrix with invariant elementary divisors by the following rule: Let the i-th column and the i-th row of a matrix correspond to the arc s_i and to the double point D_i respectively. If $\lambda(i) \neq i$ and $\lambda(i) \neq i+1$, in the row i we place

+1 in the columns i and i+1

−2 in the column $\lambda(i)$

and zero for all other elements of the row.

If $\lambda(i) = i+1$, in the row i we place

+1 in the column i

−1 in the column i+1

and zero in the remaining places of the row.

If $\lambda(i) = i$, we place

+1 in the column i+1

−1 in the column i.

This matrix has the same elementary divisors that are different from one as the matrix $(c_{\alpha\beta}^2)$.

§3. The matrix (a_{ik})

We now form matrices whose columns correspond to the bounded regions of the knot projection. We only consider two examples of this sort: Again, the i-th row corresponds to the double point D_i ($i = 1, 2, \ldots, n$), and the k-th column corresponds to the region Γ_k ($k = 1, 2, \ldots, n+1$). The elements of the matrix will be denoted by b_{ik} where i is the row index and k is the column index. We set $b_{ik} = 0$ if D_i and Γ_k are not incident. In the case that four regions (one of which may be Γ_0) come together at D_i, let $b_{ik} = +1$ or -1 depending on whether Γ_k lies to the right or to the left of the overcrossing arc $s_{\lambda(i)}$. If the regions that come together at D_i are not all mutually distinct then this means that: a region Γ_k meets itself at D_i whereas the two other regions Γ_{k_1} and Γ_{k_2} which meet at D_i are distinct. Let then $b_{ik} = 0$ and $b_{ik_\ell} = +1$ or -1 depending on whether Γ_{k_ℓ} ($\ell = 1, 2$) lies to the right or to the left of the overcrossing arc.

A matrix (b'_{ik}) is obtained analogously if in the above rule the overcrossing arc is replaced by the undercrossing arc $s_i s_{i+1}$.

Furthermore, one can form certain matrices $(b_{\alpha\beta}^h)$ whose elements correspond to the regions and the double points. These are formed analogously to the matrices $(c_{\alpha\beta}^h)$ in (5), §2, Ch. II.

Finally, we will use the black regions to define a matrix.

To a double point which is a boundary point of the black regions Γ_i and Γ_k, we assign an <u>incidence number</u>

(1) $$\eta = \pm 1, 0.$$

We define $\eta = +1$ if after the orientation of the projection plane the overcrossing arc can be rotated in a positive sense over a black region into the undercrossing arc (Fig. 20). If this rotation in the positive sense can only be carried out over a white region, then we set $\eta = -1$. If a single black region meets itself at a double point, then the point receives the incidence number 0. One sees that a double point that is incident with Γ_i and Γ_k receives the same incidence number for both regions. Thus to each double point of the projection plane there is assigned uniquely, one of the numbers $+1$, -1, 0.

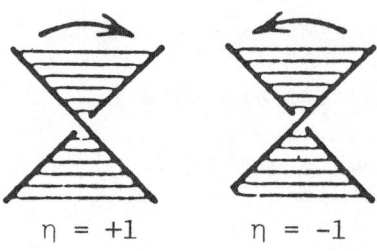

Fig. 20.

Now, let Γ_i ($i = 1, 2, \ldots, m$) be the finite black regions and let Γ_0 be the infinite (black) region. Then the following square matrix (a_{ik}) with m rows and columns is formed ([10]):

Let a_{ii} be the sum of the incidence numbers of the double points of the region Γ_i and let a_{ik} for $(i \neq k)$ $(i, k = 1, 2, \ldots, m)$ be the negative of the sum of the incidence numbers of the double points that belong simultaneously to Γ_i and Γ_k. It follows immediately from this definition that

$$a_{ik} = a_{ki}$$

and

$$a_{ii} = a_{i0} - \sum_{k \neq i} a_{ik},$$

Ch. II, §4   THE DETERMINANT OF A KNOT   33

where a_{i0} is the sum of the incidence numbers that belong simultaneously to Γ_i and to the infinite region Γ_0.

The matrix $(a'_{k\ell})$ with the elements

$$a'_{k\ell} = \sum_i b_{ik} b'_{i\ell} \quad (k,\ell = 1,2,\ldots,n+1)$$

has a simple relationship to the matrix (a_{ik}). By striking out in the matrix (a'_{ik}) the rows and columns belonging to the white regions we obtain the matrix (a_{ik});* a'_{ik} is equal to zero if Γ_i and Γ_k are colored differently.

Those elementary divisors that differ from 1 in the matrices (b_{ik}), (b'_{ik}) and (a_{ik}) are equal to the second torsion numbers. But the matrices are also of interest independently of their elementary divisors.

§4. The determinant of a knot

Among the matrices defined in the preceding section, the matrix (a_{ik}) deserves special attention. We first show that the determinant of the matrix (a_{ik}), the <u>determinant of a knot</u>, is always odd and hence nonzero.

For, if at a double point that corresponds to Γ_i and Γ_k ($i \neq k$) one changes the overcrossing edge path into an undercrossing one, then, when $k = 0$, a_{ii} changes to $a_{ii} \pm 2$ and, when $k \neq 0$,

---

*This is an error. Consider 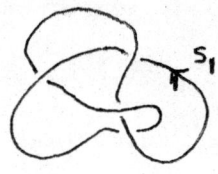 . But the result is not used subsequently.

a_{ik}, a_{ki} changes to $a_{ik} \pm 2$, i.e., the determinant of the new knot matrix is even or odd, depending on whether the original one is. By means of suitable changes of this sort, one can transform each projection into the projection of a circle. We shall assume for now that the absolute value of the determinant is an invariant, and calculate its value for the projection of the circle given in Fig. 21—it is equal to 1—thus the assertion follows.

Fig. 21.

We shall now describe and calculate the determinant in several special cases.

For alternating knots, all double points have the same incidence number with respect to black regions, since clearly the double points that are adjacent in the boundary of a black region possess the same incidence number. The incidence numbers of the points can all be taken to be, for example, equal to +1 or 0. Accordingly, the value of the determinant of the knot in this case can be written in the following form:

$$A = \begin{vmatrix} \sum_{\nu=1}^{m} d_{1\nu} & -d_{12} & \cdots & -d_{1m} \\ -d_{21} & \sum_{\nu=1}^{m} d_{2\nu} & \cdots & -d_{2m} \\ \cdots & \cdots & \cdots & \cdots \\ -d_{m1} & -d_{m2} & \cdots & \sum_{\nu=1}^{m} d_{m\nu} \end{vmatrix}.$$

Here, d_{ii} signifies the number of double points in which Γ_i and Γ_0 are contiguous, and d_{ik} ($i \neq k$) signifies the number of double points in which Γ_i and Γ_k are contiguous ($i = 1, 2, \ldots, m; k = 1, 2, \ldots, m$). The d_{ik} are thus all greater than or equal to zero.

For knots whose projections are normal, $|d_{ii}|$ is the number of double points which are incident with Γ_i and Γ_0, and the sign of d_{ii} equals the sign of the incidence numbers of these double points; $|d_{ik}|$ is the number of double points incident with Γ_i and Γ_k, and the sign of d_{ik} coincides with the sign of the incidence numbers of these double points.

For the alternating torus knots (Fig. 7) with n overcrossings, the determinant equals n (up to sign).

For the pretzel knots, depicted in Fig. 8, with a_1, a_2, a_3 overcrossings on the first, second, and third two-stringed braid parts, the determinant is equal to

$$\begin{vmatrix} a_1 + a_2 & -a_2 \\ -a_2 & a_2 + a_3 \end{vmatrix} = a_1 a_2 + a_1 a_3 + a_2 a_3.$$

§5. The invariance of the torsion numbers

We shall now show that the elementary divisors which are different from 1 of the matrices $(c_{\alpha\beta}^h)$ are knot invariants. We do this by investigating the changes produced in the matrices by the three basic operations $\Omega.1, 2, 3$. Later we shall show other ways to prove the invariance which while more natural are less elementary. These other methods make use of the group concept or of the concept of covering spaces and homology (cf. §9 and §10, Ch. III).

(Ω.1.) From a double point free arc, which with a suitable numbering we may call s_1, we form a loop. Of the various possibilities of normalization and orientation it suffices here (and also in the consideration of the other two Ω transformations), to consider only one, since then it will be clear how to proceed in the other cases.

Denote the double point of the loop D_{n+1} and the new arc s_{n+1}. Assume s_{n+1} crosses over at D_{n+1} so that, say, $\varepsilon_{n+1} = +1$. The new matrix has h more rows respectively columns than the original matrix. These h rows and h columns correspond to D_{n+1} and s_{n+1}, respectively. The rows corresponding to the other double points are unchanged except in the columns corresponding to s_1, and those columns are at most changed so that their elements corresponding to a point D_i are moved to the corresponding place in the same row of the column corresponding to s_{n+1}. Therefore if one adds the columns of index $(n+1,k)$ to those of index $(1,k)$, then one obtains a matrix which coincides with the original one up to the columns $(n+1,k)$ and the rows $(n+1,k)$ $(k = 0,1,\ldots,h-1)$. In the row $(n+1,k)$ there now appears a $+1$ in the column $(n+1,k-1)$ so that one can reduce to zero all the elements in the other rows of this column by means of successive addition or subtraction of these rows. But the matrix changed in this manner clearly has the same elementary divisors that are different from 1 as the original matrix.

(Ω.2.) As a second possibility we must consider the case where two different arcs are pushed by one another, whereby two new double points arise.

Let the overcrossing segment be numbered by s_ℓ and let the new double points be denoted D_{n+1} and D_{n+2}, and analogously number the new segments by s_{n+1} and s_{n+2} the ordering corresponding to the orientation of the knot. If, say, $\varepsilon_{n+1} = -1$, then $\varepsilon_{n+2} = +1$. The new rows $(n+1,k)$ and $(n+2,k)$ corresponding to D_{n+1} and D_{n+2} have the following property: In the columns $(n+2,k)$, all the elements are zero except for the elements in the rows $(n+1,k)$ and $(n+2,k)$, since s_{n+1} lies between the two new double points. One would obtain (except for the 2h new rows) the original columns corresponding to s_1 if the columns $(n+1,k)$ were added to the columns $(1,k)$.

In order to see that the elementary divisors different from 1 of both matrices coincide, one adds the columns $(n+2,k)$ to the columns (ℓ,k), and then one subtracts the columns $(n+2,k)$ from the columns $(\ell,k-1)$. By this the elements of the rows $(n+1,k)$ and $(n+2,k)$ in the columns (ℓ,k) are reduced to zero. By addition of the rows $(n+2,k-1)$ to the rows $(n+1,k)$ one obtains a matrix where the elements of the columns $(n+2,k)$ are 0 in all rows except for elements equal to 1 at the intersection of rows $(n+2,k)$ with columns $(n+2,k)$. All other elements in the rows $(n+2,k)$ can be removed by the addition of columns. If one then adds the columns $(n+1,k)$ to the columns $(1,k)$, then there appear in the rows $(n+1,k)$ only zeros except for elements equal to 1 at the intersection of columns $(n+1,k)$ with the rows $(n+1,k-1)$.

By addition and subtraction of the rows $(n+1,k)$ one can then, without changing the other columns, reduce all elements in the columns $(n+1,k)$ to 0 except for those in the rows $(n+1,k-1)$. The new matrix clearly has the same elementary divisors that are different from 1 as the original.

(Ω.3.) Finally, it remains to consider the operation Ω.3. For this we single out the case shown by Fig. 22.

One gets from the matrix associated with Fig. 22(a) to the matrix associated with Fig. 22(b) by first adding the rows $(\ell,k-1)$ to the rows $(n-2,k)$ and subtracting the rows $(\ell,k-1)$ from the rows $(n-1,k)$. Then add the columns $(n-1,k)$ to the columns (ℓ,k) and subtract the column $(n-1,k)$ from the columns $(\ell+1,k)$. In the cases where $s_{n-2}, s_n, s_\ell, s_{\ell+1}, s_j$ are not all different, the considerations are naturally simplified. One thus recognizes that the elementary divisors of the matrix $(c_{\alpha\beta}^h)$ are knot invariants.

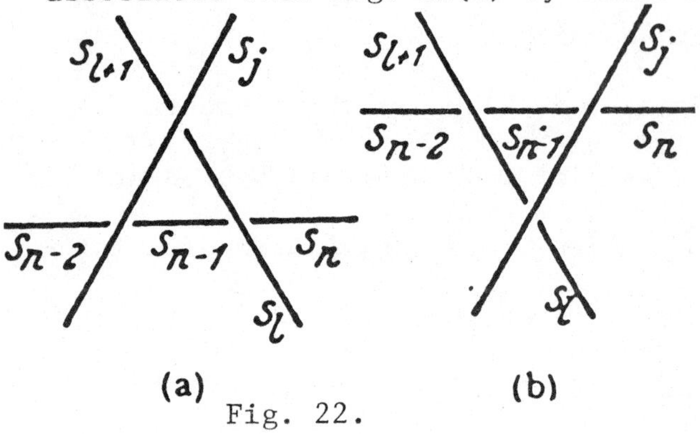

Fig. 22.

§6. The torsion numbers of particular knots

For mirror images and oppositely oriented knots, the torsion numbers are identical. One can show this directly; but it also follows from the group-theoretical interpretation of the torsion numbers (§7, Ch. III), and the relationships of the groups for mirror images and oppositely oriented knots (§5, Ch. III).

For alternating torus knots with n overcrossings, the matrix $(c_{\alpha\beta}^2)$ has only one elementary divisor different from 1, which is equal to n. Among pretzel knots, there are some that do not have second torsion numbers (cf. §11, Ch. II).

TORSION NUMBERS OF PARTICULAR KNOTS

One can classify all knots of eight or fewer crossings by using the elementary divisors of the above matrices, with h equal to 2 and 3. Among the knots with nine overcrossings on the other hand there appear knots that have the same second and third torsion numbers but can be recognized as different by other means (cf. §15, Ch. III): namely 7_4 and 9_2, 8_{14} and 9_8, and also 9_{28} and 9_{29}. Actually, the torsion numbers coincide for each pair of these knots. For them the L-polynomial matrices, defined in §14, Ch. II, are L-equivalent (§15, Ch. II, and §14, Ch. III), and the torsion numbers are determined by this equivalence class (§7, Ch. III). The following is a table of the second and third torsion numbers for the knots given in the table on pp. 126-128.

Type	$h=2$	$h=3$	Type	$h=2$	$h=3$	Type	$h=2$	$h=3$
3_{1a}	3	2,2	8_{15a}	33	16,16	9_{22a}	43	14,14
4_{1a}	5	4,4	8_{16a}	35	11,11	9_{23a}	45	22,22
5_{1a}	5	—	8_{17a}	37	13,13	9_{24a}	45	16,16
5_{2a}	7	5,5	8_{18a}	3,15	2,2,8,8	9_{25a}	47	26,26
6_{1a}	9	7,7	8_{19n}	3	4,4	9_{26a}	47	17,17
6_{2a}	11	5,5	8_{20n}	9	4,4	9_{27a}	49	19,19
6_{3a}	13	7,7	8_{21n}	15	8,8	$°9_{28a}$	51	20,20
7_{1a}	7	—	9_{1a}	9	—	$°9_{29a}$	51	20,20
7_{2a}	11	8,8	*9_{2a}	15	11,11	9_{30a}	53	22,22
7_{3a}	13	4,4	9_{3a}	19	—	9_{31a}	55	23,23
*7_{4a}	15	11,11	9_{4a}	21	7,7	9_{32a}	59	23,23
7_{5a}	17	7,7	9_{5a}	23	17,17	9_{33a}	61	25,25
7_{6a}	19	11,11	9_{6a}	27	4,4	9_{34a}	69	31,31
7_{7a}	21	13,13	9_{7a}	29	13,13	9_{35a}	3,9	20,20
8_{1a}	13	10,10	×9_{8a}	31	17,17	9_{36a}	37	10,10
8_{2a}	17	—	9_{9a}	31	5,5	9_{37a}	3,15	28,28
8_{3a}	17	13,13	9_{10a}	33	13,13	9_{38a}	57	28,28
8_{4a}	19	8,8	9_{11a}	33	7,7	9_{39a}	55	32,32
8_{5a}	21	4,4	9_{12a}	35	20,20	9_{40a}	5,15	4,4,8,8
8_{6a}	23	11,11	9_{13a}	37	16,16	9_{41a}	7,7	28,28
8_{7a}	23	5,5	9_{14a}	37	22,22	9_{42n}	7	2,2
8_{8a}	25	13,13	9_{15a}	39	23,23	9_{43n}	13	2,2
8_{9a}	25	7,7	9_{16a}	39	8,8	9_{44n}	17	10,10
8_{10a}	27	8,8	9_{17a}	39	11,11	9_{45n}	23	14,14
8_{11a}	27	14,14	9_{18a}	41	19,19	9_{46n}	3,3	7,7
8_{12a}	29	19,19	9_{19a}	41	25,25	9_{47n}	3,9	5,5
8_{13a}	29	16,16	9_{20a}	41	13,13	9_{48n}	3,9	17,17
×8_{14a}	31	17,17	9_{21a}	43	26,26	9_{49n}	5,5	10,10

---

*This table of torsion numbers differs from that given by Alexander and Briggs ([5]) for the knots 8_{12} and 9_{36}. The indices a, n mean alternating, non-alternating, respectively.

§7. The quadratic form of a knot

We have seen that the transformations of the matrices $(c_{\alpha\beta}^h)$ that are produced by the knot deformations leave unchanged those elementary divisors that are different from 1; but we have certainly not said that these elementary divisors are the only properties of those matrices that are preserved under the knot deformations. The other invariants do seem to be more difficult to comprehend, but in any case the contents differ for the various matrices given in §§2, 3, Ch. II. One result holds only for the matrices $(b_{ik}), (b'_{ik})$, and (a_{ik}). We will work with the last matrix only.

In order to examine the changes that the knot deformations make on the matrix (a_{ik}), we divide the three operations $\Omega.1,2,3$ into two classes α and β, according to the number of white or of the black regions which are changed by them ([19]).

($\Omega.1\alpha.$) Form a loop on an arc, thereby creating a new white region. Then a black region is contiguous with itself at the new double point. Hence the matrix does not change under either this operation or its inverse.

($\Omega.2\alpha.$) Two arcs are pushed past one another so that two new white regions arise. Of the two new double points, one has incidence number $+1$ and the other -1, so that the matrix also does not change under this operation or its inverse.

($\Omega.1\beta.$) The new loop of the projection encloses a black region Γ_{m+1}. The new matrix (a'_{ik}) has one more row and one more column than (a_{ik}).

In the case where Γ_{m+1} lies opposite the region Γ_0 at the new double point, then

$$a'_{m+1,m+1} = \pm 1,$$

depending on whether the double point has incidence number $+1$ or -1. Furthermore,

$$a'_{m+1,k} = a'_{k,m+1} = 0 \quad (k = 1,2,\ldots,m),$$

and for all remaining indices,

$$a'_{ik} = a_{ik}.$$

In the case where Γ_{m+1} lies opposite a region Γ_k $(k \neq 0)$ at the new double point, then we can assume without loss of generality that $\Gamma_k = \Gamma_m$. For the corresponding matrix $(a''_{\lambda k})$ we have

$$a''_{m+1,m+1} = \pm 1$$

under the same conditions as with $a'_{m+1,m+1}$ and

$$a''_{mm} = a_{mm} \pm 1, \quad a''_{m+1,m} = a''_{m,m+1} = \mp 1,$$

$$a''_{m+1,k} = a''_{k,m+1} = 0 \quad (k = 1,2,\ldots,m-1),$$

and otherwise

$$a''_{ik} = a_{ik}.$$

The nature of these changes will be clearer if we consider the quadratic form

$$f(x_1, x_2, \ldots, x_m) = \sum_{i,k} a_{ik} x_i x_k$$

which we temporarily assign to the matrix (a_{ij}). We see that the quatratic form

$$f'' = f''(x'_1, x'_2, \ldots, x'_{m+1})$$

assigned to the matrix (a''_{ik}) is transformed into the quadratic form

$$f' = f'(x_1, x_2, \ldots, x_{m+1})$$

assigned to (a'_{ik}) by the unimodular substitution

$$x'_i = x_i \quad (i = 1, 2, \ldots, m)$$

$$x'_{m+1} = x_m + x_{m+1}$$

and that

$$f'(x_1, x_2, \ldots, x_{m+1}) = f(x_1, x_2, \ldots, x_m) \pm x^2_{m+1}.$$

($\Omega.2\beta$.) Two new black regions arise by shoving two arcs past one another. In the case where the divided region is not Γ_0, we can make it Γ_m by row and column interchanges. Therefore, Γ_m is divided

Ch. II, §7   THE QUADRATIC FORM OF A KNOT   43

into Γ'_m, Γ'_{m+1}. Let Γ'_{m+2} be the new region that has both of the two new double points on its boundary. The new knot matrix (a'_{ik}) is related with (a_{ik}) by the following equations:

$$a'_{ik} = a_{ik} \quad (k,i = 1,2,\ldots,m-1)$$

$$a_{mk} = a'_{mk} + a'_{m+1,k} \quad (k = 1,2,\ldots,m-1)$$

$$a_{mm} = a'_{mm} + a'_{m+1,m+1} + 2a'_{m+1,m}$$

$$a'_{m+2,k} = a'_{k,m+2} = 0 \quad (k = 1,2,\ldots,m-1;m+2).$$

By choice of orientation we may assume that

$$a'_{m+2,m} = 1, \ a'_{m+2,m+1} = -1.$$

If

$$f' = f'(x'_1, x'_2, \ldots, x'_{m+2})$$

is the quadratic form assigned to (a'_{ik}) and if we set

$$f(x_1, x_2, \ldots, x_{m+2}) = f(x_1, x_2, \ldots, x_m) + g(x_{m+1}, x_{m+2}),$$

where

$$g(x_{m+1}, x_{m+2}) = (a'_{m+1,m+1} a'_{m+1,m+1} + 2a'_{m+1,m}) x^2_{m+1} - 2x_{m+1} x_{m+2},$$

then f' is transformed by the substitution

$$x'_i = x_i \quad (i = 1,2,\ldots,m)$$

$$x'_{m+1} = x_m + x_{m+1}$$

$$x'_{m+2} = a'_{m+1,1}x_1 + a'_{m+1,2}x_2 + \ldots + a'_{m+1,m-1}x_{m-1} + (a'_{m+1,m} + a'_{m+1,m+1})x_m$$

$$- a'_{m+1,m}x_{m+1} + x_{m+2}$$

into $f(x_1, x_2, \ldots, x_{m+2})$.

If, on the other hand, the region Γ_0 is subdivided when $\Omega.2\beta$. is applied, then the knot matrix (a''_{ik}) is related to the matrix (a_{ik}) as follows:

$$a''_{ik} = a_{ik} \quad (i,k = 1,2,\ldots,m-1)$$

$$a''_{m+2,k} = 0 \quad (k = 1,2,\ldots,m+2)$$

$$a''_{m+2,m+1} = \pm 1.$$

The values of the other $a''_{m+1,k}$ play no role; it is easy to see that the form assigned to $(a''_{\lambda k})$ is also equivalent to the form $f + g$.

($\Omega.3$.) Finally, $\Omega.3$ remains to be considered. Here, a black region is changed into a white region, or conversely (Fig. 23). Suppose first that Γ_0 does not have any part of the triangle in its boundary. Suppose further that the interior of the triangle is white; we label the neighboring black regions with $\Gamma_{m-2}, \Gamma_{m-1}, \Gamma_m$, and we again denote the knot matrix by (a_{ik}). Suppose the region Γ_{m+1} arises by the deformation, and that the new matrix is (a'_{ik}). Then

$$a'_{ik} = a_{ik} \qquad (i = 1,2,\ldots,m-3; k = 1,2,\ldots,m)$$

and furthermore

$$a'_{m+1,k} = 0 \qquad (k = 1,2,\ldots,m-3)$$

$$a'_{m+1,m-2} = a'_{m+1,m-1} = \mp 1$$

$$a'_{m+1,m} = \pm 1$$

$$a'_{m+1,m+1} = \pm 1$$

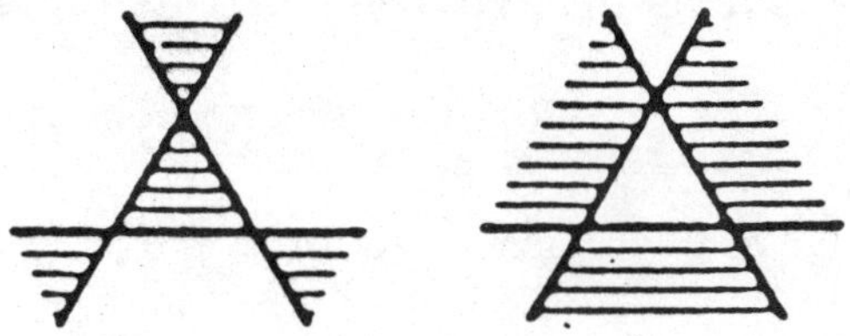

Fig. 23.

$$a'_{m-2,m-2} = a_{m-2,m-2} + a'_{m+1,m-2}$$

$$\left.\begin{array}{l} a'_{m-2,k} = a_{m-2,k} - a'_{m+1,k} \\[1em] a'_{m-1,k} = a_{m-1,k} - a'_{m+1,k} \\[1em] a'_{mk} = a_{mk} + a'_{m+1,k} \end{array}\right\} \quad (k \neq m+1).$$

The quadratic form $f' = f'(x'_1, x'_2, \ldots, x'_{m+1})$ that corresponds to (a'_{ik}) is changed by the unimodular substitution

$$x'_i = x_i$$

$$x'_{m+1} = \pm x_{m-2} \pm x_{m-1} \mp x_m \pm x_{m+1}$$

into the form

$$f(x_1, x_2, \ldots, x_m) \pm x^2_{m+1},$$

where f is the form assigned to the matrix (a_{ik}).

For the case when one of the regions that is contiguous to the triangle is Γ_0, it suffices to observe that only the $(m-1)$-st and the m-th column will be changed by the deformation. The corresponding form is in a similar way equivalent to the form $f \pm x^2_{m+1}$.

§8. Minkowski's units

A summary of the preceding section is the following.

Suppose that $f(x_1, x_2, \ldots, x_m)$ is the quadratic form assigned to the knot by means of the matrix (a_{ik}). The changes of

$$f(x_1, x_2, \ldots, x_m)$$

induced by knot deformations are compositions of the following three types of changes and their inverses:

$\Sigma.1.$ The variables of the form are unimodularly transformed:

$$x_i = \Sigma \alpha_{ik} x_k' \text{ (where the } \alpha_{ik} \text{ are integers and the determinant}$$

$$|\alpha_{ik}| = \pm 1).$$

$\Sigma.2$. The form $f(x_1, x_2, \ldots, x_m)$ is replaced by

$$f'(x_1, x_2, \ldots, x_{m+1}) = f(x_1, x_2, \ldots, x_m) \pm x_{m+1}^2.$$

$\Sigma.3$. The form $f(x_1, x_2, \ldots, x_m)$ is replaced by

$$f'(x_1, x_2, \ldots, x_{m+1}, x_{m+2}) = f(x_1, x_2, \ldots, x_m) + ax_{m+1}^2 - 2x_{m+1}x_{m+2},$$

where a is an arbitrary natural number.

Since the determinant of the matrix (a_{ik}) is not equal to zero, the form $\Sigma a_{ik} x_i x_k$ is indecomposable.

We emphasize that the invariance property of the quadratic form is not related to the group of the knot, since by §9, Ch. II, the quadratic forms for certain mirror image knots are different, while by §5, Ch. III, their groups are isomorphic. It is the only known calculable knot property that is independent from the group of the knot.

From the quadratic form one can construct knot invariants by means of the Minkowski units of quadratic forms ([24]). This is done as follows. We first define the units of a quadratic form in the following way: If p is an odd prime number and $\Sigma a_{ik} x_i x_k$ is a quadratic form with integer coefficients, whose determinant is not divisible by p, let the unit $C_p = 1$. If p is an odd prime number and $f(x_1, x_2, \ldots, x_m) = \Sigma a_{ik} x_i x_k$

is a quadratic form in which all coefficients with $i \neq k$ are divisible by p^2, then

$$(1) \quad f(x_1, x_2, \ldots, x_m) \equiv a_1 x_1^2 + a_2 x_2^2 + \ldots + a_{m-q} x_{m-q}^2 + p(a_{m-q+1} x_{m-q+1}^2 + \ldots + a_m x_m^2) \pmod{p^2}.$$

In the case that no a_i is divisible by p, then define

$$(2) \quad C_p = \left(\frac{(-1)^{[\frac{q}{2}]} a_{m-q+1} \cdots a_m}{p} \right).$$

Here,

$$\left(\frac{a}{p} \right)$$

denotes the Legendre symbol and $[a]$ denotes the greatest integer which is smaller than or equal to a. Finally, if $f'(x_1, x_2, \ldots, x_m)$ is an arbitrary form, then we may transform f' into the normal form (1) by the introduction of new variables using substitutions with rational coefficients. Then we set C_p equal to the unit (2) of (1). Minkowski showed that every form can be changed into a normal form and that the unit is well defined. From the preceding it follows that any two quadratic forms with integer coefficients which can be transformed into one another by a change of variables with rational coefficients, will have the same unit C_p. It is easy to calculate the units by using the definition given here.

Finally, we will also use the following property of C_p. Suppose that the quadratic form

$$h = h(x_1, x_2, \ldots, x_k, x_{k+1}, \ldots, x_m)$$

can be decomposed into a sum of two quadratic forms which have no variables in common,

$$h = h_1(x_1, x_2, \ldots, x_k) + h_2(x_{k+1}, \ldots, x_m).$$

Let p^{d_1} denote the highest power of an odd prime p that divides the determinant of the form h_1 and let p^{d_2} denote the highest power of the odd prime p that divides the determinant of the form h_2. Let C_{1p} and C_{2p} be the units corresponding to the forms h_1 and h_2 respectively. Then, following Minkowski ([24]), one finds that the unit C_p that corresponds to h is

$$C_p = (-1)^{\frac{p^{d_1}-1}{2} \cdot \frac{p^{d_2}-1}{2}} \cdot C_{1p} C_{2p}.$$

It follows from the results of the preceding section that those properties of quadratic forms which simultaneously belong to the forms f'', f', and f and which are preserved under unimodular substitutions of the variables are knot invariants. From this it follows directly that the units of the forms f' and f'' are equal to the units of the form f since the adjoined forms $\pm x_{m+1}^2$ and $ax_{m+1}^2 - 2x_{m+1}x_{m+2}$ possess a determinant equal to ± 1, and hence for them C_{2p} is always 1 and $d_2 = 0$. We thus obtain:

The Minkowski units C_p of the quadratic form associated with a knot matrix (a_{ik}) are knot invariants for odd primes p.

Minkowski also defined units relative to the prime 2. These are not, however, knot invariants, since they depend on the number of variables, and the number of variables is changed by knot deformations.

§9. Minkowski's units for particular knots

We will evaluate some of the Minkowski units for knots in the knot table (pp. 126-128).

Suppose that in the oriented plane, we are given the projection of the two trefoils (Fig. 2), which are mirror images of each other. The associated forms are

$$f = 3x^2 \quad \text{and} \quad f' = -3x^2,$$

both of which are normal forms for $p = 3$. We have that

$$C_3 = \left(\frac{1}{3}\right) = 1 \quad \text{and} \quad C_3' = \left(\frac{-1}{3}\right) = -1.$$

This is a simple proof of the topological difference of the two trefoils. In general, the following holds:

A knot is not isotopic to its mirror image if the highest power of some prime number of the form $4\ell+3$ that divides the determinant of the knot matrix is odd.

If the quadratic form h is a normal form associated with a knot matrix for some prime number, then $-h$ is a normal form associated with the mirror image. The number q in (1), §8, Ch. II, is odd for a prime number that divides the determinant to an odd maximal power, and thus the unit C_p' associated with the mirror image is

$$C'_p = \left(\frac{-1}{p}\right)C_p,$$

i.e.,

$$C'_p = -C_p,$$

if p is of the form $4\ell + 3$.

The necessary property of amphicheiral knots just found, that no prime number of the form $4\ell+3$ to an odd maximal power must divide the determinant is not sufficient; for, by §12, Ch. III, none of the torus knots are amphicheiral.

Below, we compile the C_p for the knots in the table on pp. 126-128. $C_p = \pm 1$ means that a knot and its mirror image yield the same C_p but with opposite sign. The knots 7_4 and 9_2 that have the same second and third torsion numbers can be differentiated by means of the C_p.

Type	c_p	Type	c_p	Type	c_p
3_{1a}	$C_3 = \pm 1$	8_{15a}	$C_3 = \pm 1; C_{11} = \pm 1$	9_{22a}	$C_{43} = \pm 1$
4_{1a}	$C_5 = -1$	8_{16a}	$C_5 = -1; C_7 = \pm 1$	9_{23a}	$C_3 = +1; C_5 = +1$
5_{1a}	$C_5 = +1$	8_{17a}	$C_{37} = -1$	9_{24a}	$C_3 = +1; C_5 = -1$
5_{2a}	$C_7 = \pm 1$	8_{18a}	$C_3 = -1; C_5 = +1$	9_{25a}	$C_{47} = \pm 1$
6_{1a}	$C_3 = +1$	8_{19n}	$C_3 = \pm 1$	9_{26a}	$C_{47} = \pm 1$
6_{2a}	$C_{11} = \pm 1$	8_{20a}	$C_3 = +1$	9_{27a}	$C_7 = +1$
6_{3a}	$C_{13} = -1$	8_{21n}	$C_3 = \pm 1; C_5 = +1$	9_{28a}	$C_3 = \pm 1; C_{17} = -1$
7_{1a}	$C_7 = \pm 1$	9_{1a}	$C_3 = +1$	9_{29a}	$C_3 = \pm 1; C_{17} = -1$
7_{2a}	$C_{11} = \pm 1$	9_{2a}	$C_3 = \pm 1; C_5 = +1$	9_{30a}	$C_{53} = -1$
7_{3a}	$C_{13} = +1$	9_{3a}	$C_{19} = \pm 1$	9_{31a}	$C_5 = +1; C_{11} = \pm 1$
7_{4a}	$C_3 = \pm 1; C_5 = -1$	9_{4a}	$C_3 = \pm 1; C_7 = \mp 1$	9_{32a}	$C_{59} = \pm 1$
7_{5a}	$C_{17} = -1$	9_{5a}	$C_{23} = \pm 1$	9_{33a}	$C_{61} = +1$
7_{6a}	$C_{19} = \pm 1$	9_{6a}	$C_3 = \pm 1$	9_{34a}	$C_3 = \pm 1; C_{23} = \pm 1$
7_{7a}	$C_3 = \pm 1; C_7 = \pm 1$	9_{7a}	$C_{29} = +1$	9_{35a}	$C_3 = \pm 1$
8_{1a}	$C_{13} = -1$	9_{8a}	$C_{31} = +1$	9_{36a}	$C_{37} = +1$
8_{2a}	$C_{17} = -1$	9_{9a}	$C_{31} = \pm 1$	9_{37a}	$C_3 = +1; C_5 = -1$
8_{3a}	$C_{17} = +1$	9_{10a}	$C_3 = \pm 1; C_{11} = +1$	9_{38a}	$C_3 = \pm 1; C_{19} = \pm 1$
8_{4a}	$C_{19} = \pm 1$	9_{11a}	$C_3 = \pm 1; C_{11} = \pm 1$	9_{39a}	$C_5 = -1; C_{11} = \pm 1$
8_{5a}	$C_3 = \pm 1; C_7 = \mp 1$	9_{12a}	$C_5 = +1; C_7 = \pm 1$	9_{40a}	$C_3 = \pm 1; C_5 = -1$
8_{6a}	$C_{23} = \pm 1$	9_{13a}	$C_{37} = +1$	9_{41a}	$C_7 = +1$
8_{7a}	$C_{23} = \pm 1$	9_{14a}	$C_{37} = -1$	9_{42n}	$C_7 = \pm 1$
8_{8a}	$C_5 = +1$	9_{15a}	$C_3 = \pm 1; C_{13} = +1$	9_{43n}	$C_{13} = +1$
8_{9a}	$C_5 = +1$	9_{16a}	$C_3 = \pm 1; C_{13} = -1$	9_{44n}	$C_{17} = +1$
8_{10a}	$C_3 = \pm 1$	9_{17a}	$C_3 = \pm 1; C_{13} = +1$	9_{45n}	$C_{23} = \pm 1$
8_{11a}	$C_3 = \pm 1$	9_{18a}	$C_{41} = -1$	9_{46n}	$C_3 = +1$
8_{12a}	$C_{29} = -1$	9_{19a}	$C_{41} = +1$	9_{47n}	$C_3 = \pm 1$
8_{13a}	$C_{29} = -1$	9_{20a}	$C_{41} = -1$	9_{48n}	$C_3 = \pm 1$
8_{14a}	$C_{31} = \pm 1$	9_{21a}	$C_{43} = \pm 1$	9_{49n}	$C_5 = -1$

*)

---

*) Y. Shinohara: <u>Note on the Minkowski unit of knots</u>. Kwansei Gakuin Univ. Annual Stud. 27 (1978), 169-171, asserts that for the knot type 9_{10a}, $C_3 = \pm 1$, $C_{11} = \pm 1$ and for the knot type 9_{33a}, $C_{61} = -1$.

§10. A determinant inequality

The matrix (a_{ik}) can be used for a fairly extensive classification of knots. To do this we shall first establish some determinant inequalities ([10], [20], [25], [31]). We consider square matrices (a_{ik}) that satisfy the condition

$$(1) \qquad a_{ii} \geq \sum_{\substack{k=1 \\ k \neq i}}^{m} |a_{ik}| \qquad (i = 1, 2, \ldots, m).$$

The <u>principal minors</u> of the determinant $A = \|a_{ik}\|$ are those subdeterminants of A which arise by striking out ℓ pairs, each pair consisting of a row and a column that intersect on the principal diagonal ($\ell < m$). The principal minors clearly still satisfy condition (1). The determinant A is defined to be irreducible if it does not decompose into a product of principal minors.

Suppose that A is irreducible and satisfies condition (1); then the following holds:

The determinant $A = \|a_{ik}\|$ is zero if and only if there exist units

$$\varepsilon_k = \pm 1 \qquad (k = 1, 2, \ldots, m)$$

such that the a_{ik} satisfy the conditions

$$(2) \qquad \sum_{k=1}^{m} a_{ik} \varepsilon_k = 0 \qquad (i = 1, 2, \ldots, m).$$

The principal minors of A with less than m rows are always positive and if $A \neq 0$ then $A > 0$ also.

Ch. II, §10 A DETERMINANT INEQUALITY

From this we obtain an estimate for A:

Let

(3)
$$s_1 = a_{11} - \sum_{k=2}^{m} |a_{1k}|$$

and let A_1 be the determinant of the matrix complementary to a_{11}; then we have for A, (expanding by the first row,)

$$A = s_1 A_1 + \bar{A},$$

where $\bar{A}$ arises from A by replacing a_{11} by $\sum_{k=2}^{m} |a_{1k}|$. $\bar{A}$ satisfies conditions (1) and is irreducible since A is. Therefore, $\bar{A} \geq 0$ and indeed if (2) does not hold for A then $\bar{A} > 0$. Accordingly,

(4)
$$A \geq s_1 A_1,$$

where the equality holds if and only if condition (2) is satisfied for $\bar{A}$.

A version of the inequality (4) holds for reducible determinants. Suppose that

$$A = A^{(1)} A^{(2)} \ldots A^{(r)},$$

where the

$$A^{(i)} \quad (i = 1, 2, \ldots, r)$$

are irreducible (call them the irreducible components of A). It follows from the application of (4) to $A^{(1)}$ that

(5)
$$A \geq s_1 A_1 \quad (A_1 \text{ is complementary to } a_{11}),$$

where the greater than sign holds if the greater than sign holds in the estimate for $A^{(1)}$ and if conditions (2) are satisfied for no $A^{(i)}$.

We now apply this determinant inequality to determinants with integer entries.

Let A be a determinant with integer entries, which satisfies condition (1), and let

$$\text{(6)} \qquad \sum_{i=1}^{m} s_i \geq 2.$$

Suppose further that the same conditions are satisfied for each principal minor, i.e., we must have for the s_i' formed from the elements of a row according to equation (3) that

$$\text{(7)} \qquad \Sigma s_i' \geq 2,$$

then ([10])

$$\text{(8)} \qquad A \geq \sum_{v=1}^{m} s_v + \sum_{u,v}^{u>v} |a_{uv}|.$$

In order to prove this, we reorder the rows and columns so that $s_1 \neq 0$ and furthermore that for each of the principal minors A_k, that arises from A by striking out the first k rows and columns, the s_{1k} formed by formula (3) is $\neq 0$. If A_k is a reducible determinant with $s_{1k} = 1$, then in the largest irreducible principal minor, which contains the row associated with s_{1k} as the first row, there is an $s_{ik} \neq 0$ with $i \neq 1$. Therefore, by (5)

$$A > A_1 \geq A_1 + 1 \quad \text{for} \quad s_1 = 1$$

and

$$A \geq s_1 A_1 \quad \text{for} \quad s_1 > 1$$

and, hence, in case $A_1 > 1$, then

$$A \geq s_1 + A_1.$$

Analogously,

$$A_1 \geq s_{11} + A_2 = s_2 + |a_{21}| + A_2,$$

when $A_2 > 1$. Therefore,

$$A \geq s_1 + s_2 + |a_{21}| + A_2, \quad \text{etc.}$$

Since, by (7), $A_{m-1} > 1$, the inequality (8) holds.

§11. Classification of alternating knots

The inequality just obtained can be applied directly to the determinant of a knot. For alternating knots, $s_1 = d_{11}$ in the determinant $\|A_{ik}\| = A$, and s_{1k} is the number of overcrossings that Γ_k has in common with the regions Γ_i ($i = 1, 2, \ldots, k-1$).

Let w be an arbitrary closed path that lies entirely in the white regions except for double points. Suppose that if w goes through one overcrossing, then it goes through at least one other overcrossing. It follows that conditions (6) and (7) of §10, Ch. II, are satisfied for the d_{ik}, and hence

(1)
$$A \geq \sum_{i \geq k} d_{ik},$$

i.e., under the given hypotheses, the determinant of the knot is greater than or equal to the number of overcrossings of the knot projection.

On the other hand, if there is a path w that contains only one double point, then we can remove the double point by means of $\Omega.4$ (see §4, Ch. I). As we previously established, the resulting projection will still be alternating. Thus one can use $\Omega.4$ to eliminate as many double points as necessary from the alternating knot projection until either we arrive at the previous case or a circle is obtained. We therefore obtain the Theorem of Bankwitz:

The minimal number of double points over all regular projections of an alternating knot is at most equal to the magnitude of the determinant of the knot ([10]).

Thus if we restrict our attention to alternating knots, then the determinant A of a knot distinguishes the knot from all except a finite number of other alternating knot types.

Since the value of the determinant is 1 for the circle, the following holds in particular:

If a given alternating projection is the projection of a circle, then all the overcrossings can be removed with the aid of $\Omega.4$.

We will now show that there exist knots that do not have a regular alternating projection. Consider the pretzel knot, on whose first two-stringed braid part there lie three double points with positive incidence number, on whose second two-stringed braid part there lie two double points with negative incidence number, and on whose third two-stringed braid part there lie seven double points with positive incidence number. For this knot, §4, Ch. II, A = +1.

If the knot could be situated as an alternating knot then all over-crossings could be twisted out in a finite number of steps, and the knot would be isotopic to a circle. That this is not the case is shown by another invariant given in §14, Ch. II; the L-polynomial of this knot is

$$L(x) = 1 - x + x^3 - x^4 + x^5 - x^6 + x^7 - x^9 + x^{10}.$$

§12. Almost alternating knots

A projection is said to be <u>almost alternating</u>* if it satisfies the following two conditions:

---

*L. Goeritz, <u>Bemerkungen zur Knotentheorie</u>, Abh. Math. Sem. Hamburg 10 (1934), 201-210, observes that this definition should be changed in order for the construction in 3, below, to again yield an almost alternating knot. He suggests the following for a definition:

A projection is <u>almost alternating</u> if and only if:

a) Two double points that are both incident with the black regions Γ_i and Γ_k ($i \neq 0$, $k \neq 0$) have the same incidence number. Let d_{ik} designate the number of double points that are incident to Γ_i and Γ_k.

b) If Γ_i has no double point in common with Γ_0, then all double points on Γ_i have the same incidence number.

c) If the region Γ_i shares with Γ_0

a_1 double points with incidence number +1
a_2 double points with incidence number -1

then $a_1 - a_2 = d_{ii}$. If $d_{ii} \neq 0$ so that there are double points other than the above that are incident with Γ_i with incidence number different from the sign of d_{ii}, then let $|d_{ii}|$ be at least twice as large as the number of double points whose incidence number is different from the sign of d_{ii}. If $d_{ii} = 0$, all other double points that are incident with Γ_i must have equal incidence numbers.

1. For each i and k any two double points which are incident with both the black region Γ_i and the black region Γ_k possess the same incidence number.

2. Condition (1), §10, Ch. II, holds for elements of the determinant of the knot.

In order to justify the terminology, we assume that the first condition holds and investigate the geometric significance of the second condition. Let d_{ik} be as introduced in §4, Ch. II, and let ε_{ik} be the incidence number of the double points that are common to Γ_i and Γ_k. Note that (1) of §10, Ch. II, asserts

$$(1) \qquad \left|\sum_{k=1}^{m} \varepsilon_{ik} d_{ik}\right| \geq \sum_{k \neq i} |d_{ik}| \quad (i = 1, 2, \ldots, m).$$

This inequality is satisfied for a region which has no double point in common with Γ_0 if and only if all double points around this region have the same incidence numbers. If this happens we shall say the knot projection lies alternatingly around this region. For a "neighboring region" of Γ_0, that is, a region that has double points in common with Γ_0, (1) is clearly satisfied if the projection is alternating around the region. Suppose, on the other hand, that this projection is not alternating around the region. Then (1) holds only if there are at most half as many double points of the region, which have an incidence number different than those points of this region incident with Γ_0, as there are double points which are incident with Γ_0.

The following reduction criterion holds for almost alternating knots:

If the elements of the i-th row of the knot determinant satisfies

$$(2) \qquad \left| \sum_{k=1}^{m} \varepsilon_{ik} d_{ik} \right| = 1,$$

then the knot can be deformed so that the region corresponding to this row vanishes and the resulting knot is almost alternating.

Suppose that (2) holds for $i = i_0$; then by (1), at most one $d_{i_0 k}$ ($k \neq i_0$) is different from zero and it equals ± 1. Hence there are only three possibilities for the $d_{i_0 k}$:

1. Suppose $\Sigma \varepsilon_{i_0 k} d_{i_0 k} = \pm 1$ and all $d_{i_0 k}$ ($k \neq i_0$) are zero. Then $\varepsilon_{i_0 i_0} d_{i_0 i_0} = \pm 1$, and Γ_{i_0} has only one double point in common with Γ_0. This double point can be removed by means of $\Omega'.1$. The modified projection is again almost alternating.

2. Suppose $\Sigma \varepsilon_{i_0 k} d_{i_0 k} = \pm 1$ and some $\varepsilon_{i_0 k_0} d_{i_0 k_0} = \pm 1$ ($k_0 \neq i_0$). Then $d_{i_0 i_0} = 0$, and therefore the only region with which Γ_{i_0} shares a double point, D, is Γ_{k_0}. Furthermore, D can be removed by an application of $\Omega'.1$. The new projection is again almost alternating. If Γ_{k_0} is not a neighboring region of Γ_0, then (1) is still satisfied since the original and hence also the new knot projection are situated alternatingly about Γ_{k_0}. If Γ_{k_0} is a neighboring region of Γ_0, then D can either have the same, or different incidence number as the double points that are incident with Γ_{k_0} and Γ_0. If the incidence number is the same, then inequality (1) is preserved since the double point of Γ_0 and Γ_{k_0} as well as the double points of Γ_{k_0} that

have incidence numbers different from the double point common to Γ_0 and Γ_k, are preserved. For unequal incidence number, inequality (1) is clearly valid after twisting out D.

3.* Suppose $\Sigma \varepsilon_{i_0 k} d_{i_0 k} = \pm 1$ and one $\varepsilon_{i_0 k_0} d_{i_0 k_0} = \mp 1$ $(k_0 \neq i_0)$. Then $\varepsilon_{i_0 i_0} d_{i_0 i_0} = \pm 2$ and Γ_{i_0} is a neighboring region of Γ_0, and indeed there are between Γ_{i_0} and Γ_0 exactly two double points with the same incidence number. Furthermore, Γ_{i_0} has one double point D in common with Γ_{k_0} that has an incidence number with the opposite sign.

We can eliminate the double point D by rotating the loop corresponding to it by 180° (Fig. 24). Then the double points that are incident with Γ_0 interchange their incidence numbers. From this it follows that the knot remains almost alternating if Γ_{k_0} is not a neighboring region of Γ_0. But this also holds when Γ_{k_0} is a neighboring region of Γ_0 and when the double points incident with Γ_{k_0} and Γ_0 have the same incidence numbers as D, and consequently also have the same incidence number as the newly adjoined double points of Γ_{k_0} and Γ_0. On the other hand, if they possess incidence numbers with sign opposite that of D, then by previous remarks the original region Γ_0 has at least two points in common with Γ_0. Hence the new region Γ_k has in common with Γ_0 four double points with

Fig. 24.

---

*L. Goeritz, <u>Bemerkungen zur Knotentheorie</u>, Abh. Math. Sem. Hamburg 10 (1934), 201-210, observes that this construction does not yield an almost alternating knot if one uses Reidemeister's definition.

incidence numbers that are pairwise of opposite sign. But one can remove these four double points by applying $\Omega.5$. twice. Then the resulting knot projection is still almost alternating since the new region Γ_{k_0} has two double points fewer in common with Γ_0 and one double point fewer with incidence number of opposite sign in common with a region different from Γ_0 than the old region Γ_{k_0} had.

Thus, one can eliminate from an almost alternating knot projection either all the double points or one arrives at a projection for which

$$(3) \qquad \left| \sum_{k=1}^{m} \varepsilon_{ik} d_{ik} \right| > 1 \quad (i = 1, 2, \ldots, m).$$

§13. Almost alternating circles

All double points can be eliminated from an almost alternating circle projection by means of the three reduction processes given in the preceding section.*

This is proved by showing that if (3), §12, Ch. II, is satisfied then the determinant A is not ± 1. We show this for an irreducible component, $A^{(1)}$. We claim that in the inequality (1) from §12, Ch. II, the greater than sign must hold for some row of $A^{(1)}$. For, if the equality sign were to always hold, then, by adding all the columns of the matrix corresponding to $A^{(1)}$ to the first column, one would obtain a determinant whose first column is divisible by 2. The

---

*L. Goeritz, <u>Bemerkungen zur Knotentheorie</u>, Abh. Math. Sem. Hamburg 10 (1934), 201-210.

determinant of the knot would therefore be even, contrary to the result at the beginning of §4, Ch. II. Hence, suppose the greater than sign holds for the first row of $A^{(1)}$; then by (4),§10, Ch. II, $|A^{(1)}| \geq |A_1^{(1)}|$, where the determinant $A_1^{(1)}$ is formed by striking out the first row and column of $A^{(1)}$. Since $A_1^{(1)}$ is irreducible, the greater than sign certainly holds in (1), §12, Ch. II, for some row of $A^{(1)}$. Continuing, we can successively apply the estimate (5) of §10, Ch. II, and finally obtain that $|A^{(1)}|$ is greater than or equal to the magnitude of one of the elements in the principal diagonal. Therefore $|A^{(1)}|$ is greater than or equal to one:

$$\left| \sum_{k=1}^{m} d_{ik} \right| > 1.$$

One can extend the classification of almost alternating knots by means of the given estimates. Namely, if one considers those almost alternating knots in which each region neighboring on Γ_0 has at least two overcrossings in common with each of the regions that are different from Γ_0 and, if one agrees further that there is no path inside the white regions that meets only one double point, then the following holds:

The minimal number of overcrossings of such knots is less than five times the knot determinant.

§14. The L-polynomial[*] of a knot

In concluding these elementary considerations, we will assign to each knot a matrix $(\ell_{ik}(x))$ whose elements are polynomials, and show that the elementary divisors of this matrix are invariants of the knot ([3], [29]).

Again, denote the n double points of the knot projection by D_i and denote the arcs between undercrossing points by s_i. Suppose that the edge path $s_i s_{i+1}$ is crossed over by $s_{\lambda(i)}$ at D_i; let $\varepsilon_i = \pm 1$ be the characteristic to D_i that was defined in (1) of §2, Ch. II.

We now form a matrix of n rows corresponding to the D_i and n columns corresponding to the s_i. The row corresponding to D_i is as follows:

Case 1. When $\varepsilon_i = +1$ and $\lambda(i) \neq i$ or $i + 1$, write x in the column corresponding to s_i, -1 in the column corresponding to s_{i+1}, and $1 - x$ in the column corresponding to $s_{\lambda(i)}$ and zero in the remaining places.

Case 2. When $\varepsilon_i = +1$ and $\lambda(i) = i$, write 1 in the column corresponding to s_i, -1 in the column corresponding to s_{i+1}, and zero in the remaining places.

Case 3. When $\varepsilon_i = +1$ and $\lambda(i) = i + 1$, write x in the column corresponding to s_i, $-x$ in the column corresponding to s_{i+1}, and zero in the remaining places.

---

[*] Commonly called the <u>normalized</u> Alexander polynomial.

Case 4. When $\varepsilon_i = -1$ and $\lambda(i) \neq i, i+1$, write 1 in the column corresponding to s_i, $-x$ in the column corresponding to s_{i+1}, $x - 1$ in the column corresponding to $s_{\lambda(i)}$, and zero in the remaining places.

Case 5. When $\varepsilon_i = -1$ and $\lambda(i) = i$, write x in the column corresponding to s_i, $-x$ in the column corresponding to s_{i+1}, and zero in the remaining places.

Case 6. When $\varepsilon_i = -1$ and $\lambda(i) = i + 1$, write 1 in the column corresponding to s_i, -1 in the column corresponding to s_{i+1}, and zero in all the remaining places.

We now consider the elements of this matrix as "L-polynomials" with integer coefficients. The totality of these polynomials

$$f(x) = \sum_{i=n}^{n+m} a_i x^i$$

(n, m, a_i are integers and $m \geq 0$; note that n can also be negative) form an integral domain whose "units" are the polynomials $\pm x^n$. Here, an element $f(x)$ is called a unit if the multiplicative inverse, $(f(x))^{-1}$, also belongs to the integral domain.

It can be shown that the elementary divisors of the matrix $(\ell_{ik}(x))$ are knot invariants up to unit multiples. To prove this, we again consider how the matrix is modified by the three operations $\Omega.1, 2, 3$. We will thus consider the same transformations as in §5, Ch. II, and retain the notation given there.

($\Omega.1$.) The row corresponding to the new D_{n+1} contains in the columns corresponding to s_{n+1} and s_1 the elements x and $-x$ respectively, while the remaining elements are equal to zero. Therefore, if we add the column corresponding to s_{n+1} to the column corresponding to s_1, then in the row assigned to D_{n+1} the only nonzero element is x in the column corresponding to s_{n+1}. By successive addition of $\pm x^{\ell}$ ($\ell = 0, +1, -1$) times this row, one can make all the other elements zero in the column corresponding to s_{n+1}.

($\Omega.2.$) The new rows have the form:

	$s_1, \ldots, s_{\ell-1}, s_\ell$	$s_{\ell+1}, \ldots, s_{n-1}, s_n$	s_{n+1}	s_{n+2}
D_{n+1}	$0, \ldots, 0, x-1,$	$0, \ldots, 0, 0,$	$1,$	$-x$
D_{n+2}	$-1, \ldots, 0, 1-x,$	$0, \ldots, 0, 0,$	$0,$	x

Otherwise, there are only zeros in the column corresponding to s_{n+2}. The original column corresponding to s_1 arises by addition of the new column corresponding to s_{n+1} to the new column corresponding to s_1.

In order to see that the elementary divisors change at most by a unit multiple, one first adds the row corresponding to D_{n+2} to the row corresponding to D_{n+1}. Now one adds the column corresponding to s_{n+1} to the column corresponding to s_1. Then by multiplying the column corresponding to s_{n+2} by the appropriate units and adding to the column corresponding to s_1 and s_ℓ, we obtain a matrix whose row corresponding to D_{n+2} has all zeros except for an x on the diagonal. Similarly, by adding unit multiples of the row corresponding to D_{n+1} to the other rows we get the matrix

$$\begin{pmatrix} \left(\ell_{ik}(x)\right) & \begin{matrix}0 & 0\\ \vdots & \vdots\end{matrix} \\ 0\ldots\ldots\ldots 0 \quad 1 \quad 0 \\ 0\ldots\ldots\ldots\ldots\ldots\ldots x \end{pmatrix}.$$

($\Omega.3.$) The submatrices which correspond to the points and arcs that are affected have the form (keep in mind the immediately following figures):

(1)

	s_j	s_ℓ	$s_{\ell+1}$	s_{n-2}	s_{n-1}	s_n
D_ℓ	$x-1,$	$1,$	$-x,$	$0,$	$0,$	0
D_{n-2}	$x-1,$	$0,$	$0,$	$x,$	$-1,$	0
D_{n-1}	$0,$	$1-x,$	$0,$	$0,$	$x,$	-1

(2)

	s_j	s_ℓ	$s_{\ell+1}$	s_{n-2}	s_{n-1}	s_n
D_ℓ	$x-1,$	$1,$	$-x,$	$0,$	$0,$	0
D_{n-2}	$0,$	$0,$	$1-x,$	$x,$	$-1,$	0
D_{n-1}	$1-x,$	$0,$	$0,$	$0,$	$x,$	-1

The remaining elements of the corresponding rows and of the column corresponding to s_{n-1} are equal to zero. In order to proceed from (1) to (2), one adds the row corresponding to D_ℓ in (1) to the row corresponding to D_{n-2} and subtracts the row corresponding to D_ℓ from the row corresponding to D_{n-1}. Then one adds the column corresponding to s_{n-1} to the column corresponding to s_ℓ and subtracts the column corresponding to s_{n-1} from the column corresponding to $s_{\ell+1}$.

We can sharpen this result still further as follows: The modifications of the matrix $(\ell_{ik}(x))$ induced by the operations Ω and Ω' are consequences of the following elementary matrix transformations:

$\Sigma.\xi.1.$ Interchange rows (columns.)

$\Sigma.\xi.2.$ Multiply all of the elements of a row (column) by $\pm x$.

$\Sigma.\xi.3.$ Add a row (column) to another.

$\Sigma.\xi.4.$ Adjoin or delete a row all of whose elements are zero.

$\Sigma.\xi.5.$ Adjoin or delete simultaneously a row and a column where the element that belongs to both this row and to this column equals one, and all other elements of the row and column are equal to zero.

Matrices that arise from one another by means of the transformations $\Sigma.\xi.$ are said to be <u>L-equivalent</u>. L-equivalent matrices clearly have elementary divisors that differ by at most a factor of $\pm x^n$ (the converse however does not hold). Hence the L-equivalence class of the matrix $(\ell_{ik}(x))$ is another knot invariant.

If one deletes an arbitrary column of $(\ell_{ik}(x))$, one obtains a matrix with the same elementary divisors; for, the sum of the elements of each row equals zero. Further, an arbitrary row can also be removed without changing the elementary divisors. This will be shown in §6, Ch. III.

All of the possible new matrices that arise by these deletions have determinants different from zero since the determinants arising when $x = 1$ are equal to ± 1; these determinants are only determined up to a factor of $\pm x^n$. By multiplying the determinant by a suitable factor, $\pm x^n$, we get the uniquely determined form

$$(3) \qquad L(x) = \ell_0 + \ell_1 x + \ldots + \ell_g x^g$$

with $\ell_0 > 0$, $g \geq 0$. (3) is called the <u>L-polynomial of the knot</u>.*

One can also define analogous matrices of polynomials, having the same elementary divisors, by taking as starting point the bounding relations between points and regions. This will follow directly from the group-theoretical interpretation of our matrix ([3]). We will show in §7, Ch. III, that all the torsion numbers are determined by the L-equivalence class of the matrix $(\ell_{ik}(x))$.

§15. L-polynomials of particular knots

The elementary divisors $e(x)$ of the matrix $(\ell_{ik}(x))$ of a knot do not change when we replace the knot by its mirror image, or oppositely direct the knot. One can see this directly, by deriving it using the group-theoretic interpretation of the matrix $(\ell_{ik}(x))$, or by noting the behavior of the groups with regard to mirror images and change of the sense of traversal of the knot (cf.§6 and §5, Ch. III). Changing from the knot to

---

*This is frequently called the <u>Alexander polynomial</u> of the knot.

the oppositely directed knot corresponds to the interchange of x and x^{-1}. From this one obtains the following symmetry property for L-polynomials:

$$\ell_i = \ell_{g-i} \quad (i = 0, 1, \ldots, [\tfrac{g}{2}]).$$

(Here, [a] is the greatest integer that is less than or equal to a.) The analogous situation holds for the elementary divisors. Since $\Sigma \ell_i = \pm 1$, g must be even and $\ell_{[\tfrac{g}{2}]}$ must be odd.

For the calculation of the L-polynomials for parallel knots and cable knots, and for the classification of similarly twisted cable knots using L-polynomials, see §13, Ch. III.

In the following table of L-polynomials for all the knots given in the knot table on pp. 126-128 we use the following abbreviation ([3]). The symbol $5-14+19$ signifies the L-polynomial

$$L(x) = 5 - 14x + 19x^2 - 14x^3 + 5x^4.$$

In the table* there are five pairs of knots with the same L-polynomial. Each of the two starred pairs have, by §6, Ch. II, different second and third torsion numbers, and their matrices $(\ell_{ik}(x))$ are not L-equivalent. On the other hand, for the three

---

*The table differs from the one given by Alexander ([3]) for the knot 9_{36}.

remaining non-isotopic (by §9, Ch. II; §§14, 15, Ch. III) pairs, the matrices $(\ell_{ik}(x))$ are L-equivalent (cf. §14, Ch. III).

Type	L-Polynomial	Type	L-Polynomial	Type	L-Polynomial
3_{1a}	$1 - 1$	8_{8a}	$2 - 6 + 9$	8_{7a}	$1 - 3 + 5 - 5$
4_{1a}	$1 - 3$	8_{11a}	$2 - 7 + 9$	8_{9a}	$1 - 3 + 5 - 7$
5_{2a}	$2 - 3$	8_{13a}	$2 - 7 + 11$	8_{10a}	$1 - 3 + 6 - 7$
6_{1a} } $2 - 5*$		8_{14a} } $2 - 8 + 11$		9_{48n}	$1 - 4 + 6 - 5$†
9_{46n}		9_{8a}		8_{16a}	$1 - 4 + 8 - 9$
7_{2a}	$3 - 5$	9_{12a}	$2 - 9 + 13$	8_{17a}	$1 - 4 + 8 - 11$
8_{1a}	$3 - 7$	9_{14a}	$2 - 9 + 15$	9_{11a}	$1 - 5 + 7 - 7$
7_{4a} } $4 - 7$		9_{15a}	$2 - 10 + 15$	9_{16a}	$1 - 5 + 8 - 9$
9_{2a}		9_{19a}	$2 - 10 + 17$	9_{17a}	$1 - 5 + 9 - 9$
8_{3a}	$4 - 9$	9_{21a}	$2 - 11 + 17$	9_{20a}	$1 - 5 + 9 - 11$
9_{5a}	$6 - 11$	9_{37a}	$2 - 11 + 19$†	9_{22a}	$1 - 5 + 10 - 11$
9_{35a}	$7 - 13$	9_{4a}	$3 - 5 + 11$†	8_{18a} } $1 - 5 + 10 - 13*$
5_{1a}	$1 - 1 + 1$	9_{49n}	$3 - 6 + 7$	9_{24a}	
9_{42n}	$1 - 2 + 1$	9_{7a}	$3 - 7 + 9$	9_{26a}	$1 - 5 + 11 - 13$
8_{20n}	$1 - 2 + 3$	8_{15a}	$3 - 8 + 11$	9_{27a}	$1 - 5 + 11 - 15$
6_{2a}	$1 - 3 + 3$	9_{25a}	$3 - 12 + 17$	9_{28a} } $1 - 5 + 12 - 15$
6_{3a}	$1 - 3 + 5$	9_{41a}	$3 - 12 + 19$	9_{29a}	
8_{21n}	$1 - 4 + 5$	9_{39a}	$3 - 14 + 21$	9_{30a}	$1 - 5 + 12 - 17$
9_{44n}	$1 - 4 + 7$	9_{10a}	$4 - 8 + 9$	9_{31a}	$1 - 5 + 13 - 17$
7_{6a}	$1 - 5 + 7$	9_{13a}	$4 - 9 + 11$	9_{32a}	$1 - 6 + 14 - 17$
7_{7a}	$1 - 5 + 9$	9_{18a}	$4 - 10 + 13$	9_{33a}	$1 - 6 + 14 - 19$
9_{45n}	$1 - 6 + 9$	9_{23a}	$4 - 11 + 15$	9_{34a}	$1 - 6 + 16 - 23$
9_{47n}	$1 - 7 + 11$†	9_{38a}	$5 - 14 + 19$	9_{40a}	$1 - 7 + 18 - 23$
8_{12a}	$1 - 7 + 13$	8_{19n}	$1 - 1 + 0 + 1$	9_{3a}	$2 - 3 + 3 - 3$
7_{3a}	$2 - 3 + 3$	7_{1a}	$1 - 1 + 1 - 1$	9_{6a}	$2 - 4 + 5 - 5$
7_{5a}	$2 - 4 + 5$	9_{43n}	$1 - 3 + 2 - 1$	9_{9a}	$2 - 4 + 6 - 7$
8_{4a}	$2 - 5 + 5$	8_{2a}	$1 - 3 + 3 - 3$	9_{16a}	$2 - 5 + 8 - 9$
8_{6a}	$2 - 6 + 7$	8_{5a}	$1 - 3 + 4 - 5$	9_{1a}	$1 - 1 + 1 - 1 + 1$

---

*See note on preceding page.

†The polynomial for 9_{4a} should be 3−5+5. In addition, the polynomials for 9_{47n} and 9_{48n} have been interchanged. 9_{47n} should read 1−4+6−5 and 9_{48n} should read 1−7+11.

CHAPTER III

KNOTS AND GROUPS

§1. Equivalence of braids

The problem of deciding whether two knot projections correspond to the same knot is similar to the word problem or the transformation problem for a group with generators and defining relations. For braids this similarity can be made precise by using a particular kind of projection and a particular type of deformation.

A <u>deformation</u> of an open braid is understood to be a finite sequence of operations in 3-space of the following type:

$\Delta.\zeta$. Let PQ be the segment of the braid that goes from P to Q and let PR and RQ be two segments which go from P to R and from R to Q, respectively. Suppose that the triangular area PQR does not intersect the braid except along the segment PQ. The segment PQ is replaced by PR and RQ if the figure obtained in this way is again an open braid.

$\Delta'.\zeta$. is the operation inverse to $\Delta.\zeta$.

Two braids which arise from one another by a deformation are called <u>equivalent</u>.

After normalizing the braid projection one can again translate the 3-space operations $\Delta.\zeta$ and $\Delta'.\zeta$, into operations of the projection. In addition to the projection of this operation $\Delta.\zeta.\pi.$, whose application yields a projected braid (Fig. 25), some of the operations $\Omega.2,3$

can be applied to a braid projection. These will be denoted by $\Omega.\zeta.2.$ (Fig. 26) and $\Omega.\zeta.3.$ (Fig. 27). Thus we are retaining those operations $\Omega.2.3$ which when applied to a braid projection yield a braid projection.

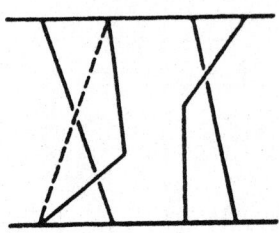 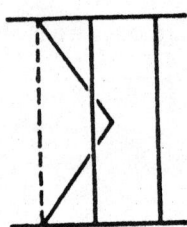 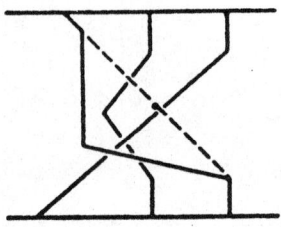

Fig. 25. $\Delta.\pi.\zeta.$ Fig. 26. $\Omega.\zeta.2.$ Fig. 27. $\Omega.\zeta.3.$

The proof that every braid deformation $\Delta.\zeta.$ can be decomposed into a finite sequence of the operations $\Omega.\zeta.2,3$ and their inverses is similar to the proof of the analogous assertion for knots which was given in §3, Ch. I.

Given two q-stringed braids z_1 and z_2, one can form a new braid $z_3 = z_1 z_2$ by means of hanging z_2 onto z_1, i.e., to z_1 there corresponds a rectangle with the opposite sides g_{11}, g_{21} and the point series A_{i1} and B_{i1}; and to z_2 there corresponds a rectangle with the opposite sides g_{12}, g_{22} and the point series A_{i2}, B_{i2} ($i = 1,2,\ldots,q$). The rectangles are now juxtaposed so that an affine transformation of the second rectangle will make B_{i1} coincide with A_{i2}; finally the segment $g_{21} = g_{12}$ is erased. The new figure is again a q-stringed braid. Let s_i denote the braid with only one double point, at which the i-th string of the braid crosses over the (i+1)-st ($i=1,2,\ldots,q-1$), and let s_i^{-1} denote the braid in which the i-th string crosses under the (i+1)-st; then one may observe that each braid with d double points can be uniquely represented as $z = s_{\alpha_1}^{\varepsilon_1} s_{\alpha_2}^{\varepsilon_2} \ldots s_{\alpha_d}^{\varepsilon_d}$ ($\varepsilon_i = \pm 1$).

Clearly, if z_1 and $\bar{z}_1$ are equivalent, and z_2 and $\bar{z}_2$ are equivalent, then $z_1 z_2$ and $\bar{z}_1 \bar{z}_2$ are also equivalent.

§2. The braid group

We now form the group, Z_q, of the q-stringed braids ([7]). As <u>group elements</u> we take the <u>classes of equivalent braids</u>, $[z]$. The <u>product</u> of two classes $[z_1]$ and $[z_2]$ is defined by

$$[z_1] \cdot [z_2] = [z_1 z_2].$$

The product is well defined and is associative. The braid without crossings represents the unit element. Furthermore, if

$$z = s_{\alpha_1}^{\varepsilon_1} s_{\alpha_2}^{\varepsilon_2} \ldots s_{\alpha_d}^{\varepsilon_d},$$

then $[s_{\alpha_d}^{-\varepsilon_d} \ldots s_{\alpha_2}^{-\varepsilon_2} s_{\alpha_1}^{-\varepsilon_1}]$ is the element inverse to $[z]$.

Since $[z] = [s_{\alpha_1}]^{\varepsilon_1} [s_{\alpha_2}]^{\varepsilon_2} \ldots [s_{\alpha_d}]^{\varepsilon_d}$, the elements

(1) $$S_i = [s_i] \quad (i = 1, 2, \ldots, q-1)$$

form a system of generators of the braid group Z_q.

One finds the defining relations of the braid group Z_q by investigating the changes in the word $W(S_i) = S_{\alpha_1}^{\varepsilon_1} S_{\alpha_2}^{\varepsilon_2} \ldots S_{\alpha_d}^{\varepsilon_d}$ that are induced by the braid operations.

The operation $\Omega.\zeta.2$ (Fig. 26) and its inverse induce the insertion or deletion, respectively, of a factor $S_i^{\varepsilon} S_i^{-\varepsilon}$ ($\varepsilon = \pm 1$), which can be considered as a relation in the free group of the S_i.

The operation $\Delta.\pi.\zeta.$ (Fig. 25) effects at most a redistribution of the double points. The simplest redistribution gives a permutation of two consecutive elements, $S_{\alpha_i}^{\varepsilon_i}, S_{\alpha_{i+1}}^{\varepsilon_{i+1}}$, in a word $W(S_i)$. In this, clearly $\alpha_i \neq \alpha_{i+1} \pm 1$. This yields the relations

(2) $$S_i S_k = S_k S_i \quad (k \neq i-1, i+1).$$

$\Omega.\zeta.3.$ (Fig. 27) involves three neighboring strings, for instance, the strings i, $i+1$, $i+2$; then we obtain the relations

$$S_i^\varepsilon S_{i+1}^\varepsilon S_i^{\pm 1} = S_{i+1}^{\pm 1} S_i^\varepsilon S_{i+1}^\varepsilon \quad (\varepsilon = \pm 1).$$

For all possible choices of exponents, these can be rewritten as

(3) $$S_i S_{i+1} S_i = S_{i+1} S_i S_{i+1} \quad (i = 1, 2, \ldots, q-2).$$

Therefore (2) and (3) are the defining relations of Z_q in terms of the generators (1).

The question of whether two open braids are equivalent is thus reduced to the so-called word problem for the group Z_q with respect to the generators (1), i.e., to the problem of deciding whether two words $W(S)$ and $W'(S)$ determine the same group element in the group with the generators (1) and the defining relations (2) and (3). The question of deciding when two closed braids are equivalent can be reduced to the transformation problem in the braid group. Namely, one sees that the closed braids that are assigned to the open braids $s_i^\varepsilon z s_i^{-\varepsilon}$ and z are equivalent, and conversely, that two open braids z and z' are related by $[z] = [z_1][z'][z_1]^{-1}$ if they correspond to the same closed braid.

The word problem is solved for all Z_q; in contrast, the transformation problem is not. Z_2 is an infinite cyclic group. Z_3 is isomorphic to the group of the trefoil knot ([7]). The transformation problem can be solved in Z_3 (§12, Ch. III).

Those braids in which A_i is always connected with B_i give rise to a noteworthy subgroup I_q of Z_q. Numerous new braid properties can be recognized by displaying the generators and defining relations of I_q. Using these properties one can completely classify the braids corresponding to the cable knots ([13]).

§3. Definition of the group of a knot

Now we again focus on the problem formulated in §2, Ch. II: to produce computable knot invariants, and to describe how one can obtain a group with generators and defining relations from the knot projection ([28]).

The regular normalized projection of a knot with n double points $D_1, D_2, \ldots, D_n$ will be subdivided into n arcs $s_1, s_2, \ldots, s_n$, each of which leads from an undercrossing place to another. Assume that a positive sense for running around the knot and a positive sense of rotation in the projection plane have been established. We also assume that the indexing is as described in §2, Ch. II. To the n segments s_i we formally assign n generators

(1) $$s_1, s_2, \ldots, s_n$$

and to each double point D_i, at which $s_{\lambda(i)}$ crosses over the segment $s_i s_{i+1}$, we assign the relation

(2) $$R_i(S) = s_{i+1}^{-1} s_{\lambda(i)}^{\varepsilon_i} s_i s_{\lambda(i)}^{-\varepsilon_i} \quad (i = 1, 2, \ldots, n)$$

where, as before, ε_i equals $+1$ or -1, depending on whether or not the direction of $s_{\lambda(i)}$ can be rotated into the direction of s_i by a positive angle α, where $\alpha < 180°$. In (2), S_{n+1} is defined to be S_1. The group $\mathcal{W}$ determined by the generators (1) and the defining relations (2) is called the <u>group of the knot</u>. Similarly, one can also define the group of a link.

It follows from the defining relations (2) that <u>the quotient group of the knot group with respect to its commutator subgroup is an infinite cyclic group</u>. That is, if the generators are permitted to commute, then the relations (2) state that

$$s_1 = s_2 = \ldots = s_n.$$

The group of a circle is the free group with one generator. That follows from the proof of invariance in §4, Ch. III, and the calculation of the group using a suitable projection.

If one takes the group of a link of r polygons, and forms the quotient group with the commutator group, then one obtains a free Abelian group with r generators. The group of two unlinked polygons is the free product of their groups; in particular, the group of two unlinked circles is the free group with two free generators.

The group W of the link formed by two polygons $k^{(1)}$ and $k^{(2)}$ can be used to compute the intertwining number of those polygons, defined in §1, Ch. II. The second commutator subgroup of a group G is the subgroup generated by the elements

$$K_2 = SK_1 S^{-1} K_1^{-1}$$

where S is an arbitrary element of G and K_1 is an arbitrary element of the commutator subgroup of G. Now, if K_2 is the second commutator group of W and $F = W/K_2$ is the quotient group of W modulo K_2, and furthermore, if $S^{(1)}$ and $S^{(2)}$ are two generators of W, which correspond to arcs of $k^{(1)}$ and $k^{(2)}$, respectively, then the order of the commutator

$$S^{(1)} S^{(2)} S^{(1)-1} S^{(2)-1}$$

in F is equal to the intertwining number of $k^{(1)}$ and $k^{(2)}$.

There are many natural ways to obtain generators and relations. For example, consider a projection of a closed braid that corresponds to an open braid consisting of q strings, which connect the points $A_1, A_2, \ldots, A_q$ respectively with the points $B_{k_1}, B_{k_2}, \ldots, B_{k_q}$. If $S_{A_1}, S_{A_2}, \ldots, S_{A_q}$ are generators which correspond to the q arcs that contain the A_i and B_{k_i}, then we can express successively all remaining generators in terms of the S_{A_i} by means of the relations. Then there remain exactly q relations which are of the form

$$(3) \qquad S_{A_i}^{-1} L_i S_{A_{k_i}} L_i^{-1} \qquad (i = 1, 2, \ldots, q),$$

where the L_i are particular powers of products of the generators which satisfy the identity

$$(4) \qquad \prod_{i=1}^{q} L_i S_{A_{k_i}} L_i^{-1} = \prod_{i=1}^{q} S_{A_i}$$

in the free group of the S_{A_i}. Relations (3) and (4) can be used for an algebraic characterization of link groups (which include knot groups). Since every link can be deformed into a braid, every link group can be rewritten in the form characterized by (3) and (4). Conversely, corresponding to any relation system (3) which satisfies relation (4), one can construct a closed braid to which there is assigned a group by means of the algorithm described at the beginning of this section. The group so obtained is isomorphic ([7]) to the group with the generators S_{A_i} ($i = 1, 2, \ldots, q$) and the relations (3).

Another algorithm for the formation of a knot group can be given in the following way. To each bounded region Γ_k ($k = 1, 2, \ldots, n+1$) of the projection plane assign a generator

$$(5) \qquad T_k \qquad (k = 1, 2, \ldots, n+1),$$

and let T_0 (which is associated with the unbounded region Γ_0) represent the unit element. Let a relation $R_i(T)$ ($i = 1,2,\ldots,n$) be assigned to each double point D_i as follows. If, for instance, $\Gamma_{\mu_\ell(i)}$ ($\ell = 1,2,3,4$) are the four regions arranged clockwise around D_i (we permit $\Gamma_{\mu_\ell(i)}$ to be Γ_0) and if $\Gamma_{\mu_1(i)}, \Gamma_{\mu_2(i)}$ as well as $\Gamma_{\mu_3(i)}, \Gamma_{\mu_4(i)}$ are adjacent along the two undercrossing arcs at the double point D_i (therefore $\Gamma_{\mu_2(i)}, \Gamma_{\mu_3(i)}$ as well as $\Gamma_{\mu_4(i)}, \Gamma_{\mu_1(i)}$ are adjacent along the overcrossing arcs at the double point D_i), then let

(6) $$R_i'(T) = T_{\mu_1(i)}^{-1} T_{\mu_2(i)} T_{\mu_3(i)}^{-1} T_{\mu_4(i)} \quad (i = 1,2,\ldots,n).$$

As we shall see in §9, Ch. III, the group defined by (5) and (6) is isomorphic to the group defined by (1) and (2). This can also be verified directly by changing the generators and relations.

§4. Invariance of the knot group

The group defined in §3, Ch. III, by means of the generators (1) and relations (2) is a knot property of the regular normalized projection.

As in ([28]), we prove this by investigating how the generators and relations change upon application of the operations $\Omega.1,2,3$. The generators which arise from the altered projection are designated by a prime ':

($\Omega.1.$) Suppose that a loop is formed in the edge path s_n. The relation

$$s_{n+1}'^{-1} s_n'^{\varepsilon_{n+1}} s_n' s_n'^{-\varepsilon_{n+1}} = 1 \quad \text{or} \quad s_{n+1}'^{-1} s_{n+1}'^{\varepsilon_{n+1}} s_n' s_{n+1}'^{-\varepsilon_{n+1}} = 1$$

corresponds to the new double point. In both cases it follows that $S'_{n+1} = S'_n$. If we replace S'_{n+1} by S'_n everywhere, then the same relations hold among the S'_i as among the S_i.

(Ω.2.) Assume the new double points arise by shoving s_ℓ over s_n. It follows from the two relations corresponding to the new double points, namely,

$$S'^{-1}_{n+1} S'^{\varepsilon_{n+1}}_\ell S'_n S'^{-\varepsilon_{n+1}}_\ell = 1$$

$$S'^{-1}_{n+2} S'^{\varepsilon_{n+2}}_\ell S'_{n+1} S'^{-\varepsilon_{n+2}}_\ell = 1 \qquad (\varepsilon_{n+1} = -\varepsilon_{n+2} = \varepsilon)$$

that $S'_{n+2} = S'_n$. If we now set

$$S''_i = S_i \quad (i = 1,2,\ldots,n), \quad T = S'_{n+1}, \quad S''_n = S'_{n+2},$$

then there hold among the S''_i the same relations as among the S_i, and further between S''_i and T the following two relations hold:

$$T^{-1} S''^{\varepsilon}_\ell S''_n S''^{-\varepsilon}_\ell = 1, \quad S''^{-1}_n S''^{-\varepsilon}_\ell T S''^{\varepsilon}_\ell = 1.$$

the second relation is a consequence of the first.

(Ω.3.) For this operation, we have different cases to consider, each depending on the orientation which the edges of the triangle inherit from the orientation of the knot. We focus on the case represented in the figure:

Ch. III, §4 INVARIANCE OF THE KNOT GROUP

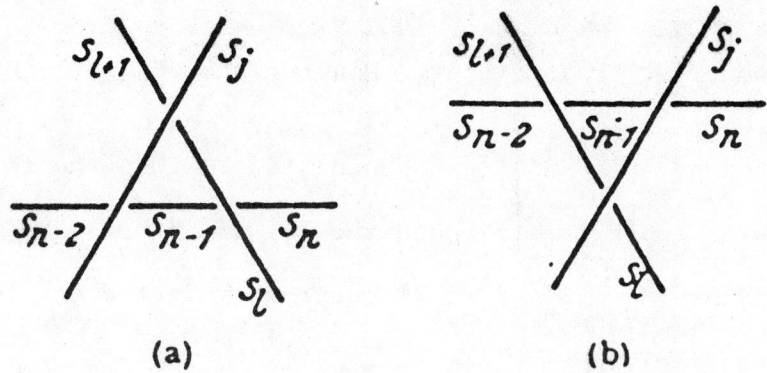

(a) (b)

The relations of the initial projection relative to the double points of the triangle are:

$$R_\ell = s_{\ell+1}^{-1} s_j^{\varepsilon_\ell} s_\ell s_j^{-\varepsilon_\ell},$$

$$R_{n-2} = s_{n-1}^{-1} s_j^{\varepsilon_{n-2}} s_{n-2} s_j^{-\varepsilon_{n-2}},$$

$$R_{n-1} = s_n^{-1} s_\ell^{\varepsilon_{n-1}} s_{n-1} s_\ell^{-\varepsilon_{n-1}}.$$

The relations after the transformation are:

$$R'_\ell = {s'}_{\ell+1}^{-1} {s'}_j^{\varepsilon'_\ell} s'_\ell {s'}_j^{-\varepsilon'_\ell},$$

$$R'_{n-2} = {s'}_{n-1}^{-1} {s'}_{\ell+1}^{\varepsilon'_{n-2}} s'_{n-2} {s'}_{\ell+1}^{-\varepsilon'_{n-2}},$$

$$R'_{n-1} = {s'}_n^{-1} {s'}_j^{\varepsilon'_{n-1}} s'_{n-1} {s'}_j^{-\varepsilon'_{n-1}}.$$

In this, $\varepsilon'_\ell = \varepsilon_\ell$, $\varepsilon'_{n-2} = \varepsilon_{n-1}$, $\varepsilon'_{n-1} = \varepsilon_{n-2}$ and furthermore $\varepsilon'_\ell = -\varepsilon_{n-1} = -\varepsilon_{n-2}$. We eliminate S_{n-1} from R_{n-1} using R_{n-2} and then eliminate S'_{n-1} from R'_{n-1} using R'_{n-2}; Now replace $S'_{\ell+1}$ in the last relation obtained by $S'^{\varepsilon'_\ell}_j S'_\ell S'^{-\varepsilon'_\ell}_j$ using R'_ℓ. Then the relations R_k and R'_k ($k \neq n-2$) among the S_i and the S'_i ($i \neq n-1$) respectively are the same, and S_{n-1}, S'_{n-1} still occur only in the one relation R_{n-2} or R'_{n-2} respectively.

Summarizing, the new relations arise from the initial ones by the addition and removal of redundant relations or by the addition and removal of generators each of which is expressible as a word in the remaining generators, via the relations. Consequently the groups defined by means of the relations are isomorphic to one another.

Besides the group itself which is invariant under Ω, Ω', there are other invariants that may be defined from the S_k of (1), §3, Ch. III, and the R_i of (2), §3, Ch. III. By varying the projections, for instance, one merely introduces new generators by means of the equations

(1) $$S_a = S_b^\varepsilon S_c S_b^{-\varepsilon}.$$

Therefore, each property that is retained after the reductions and extensions of the defining relations and after the reductions and extensions of the generators of the form (1) is a knot property.

Further, we see that each generator S_i can be represented as a conjugate of any of the others. For example,

(2) $$S_i = L_i S_1 L_i^{-1} \quad \text{where} \quad L_i = S_{\lambda(i-1)}^{\varepsilon_{i-1}} S_{\lambda(i-2)}^{\varepsilon_{i-2}} \cdots S_{\lambda(1)}^{\varepsilon_1} \quad (i=2,3,\ldots,n).$$

Therefore, a new invariant of the knot group is obtained if $S_1^h = 1$ is adjoined to the relations $R_k(S)$.

From the same considerations it can be concluded that for each element WS_1W^{-1} (W arbitrary) there is a class of elements associated with it,

$$(3) \qquad WL_{n+1}^{\varepsilon} S_1^{\ell} W^{-1} \qquad (\ell = 0, \pm 1, \pm 2, \ldots; \varepsilon = \pm 1),$$

where L_{n+1} is determined by (2) for $i = n+1$. Hence by (2),

$$S_1 = L_{n+1} S_1 L_{n+1}^{-1}.$$

That these classes are knot invariants can easily be verified.

§5. The group of the inverse knot and of the mirror image knot

The group of a knot is isomorphic to the group of the inverse knot and to the group of the mirror image knot ([9]); these knots can therefore not be distinguished by means of the structure of their groups. Starting with the description of the group given by (1) and (2) of §3, Ch. III, we reverse the sense of direction of the knot projection and now assign a generator S_i' to the arc which corresponded to S_i. Among the S_i' the following relations hold:

$$(1) \qquad S_{i+1}'^{-1} S_{\lambda(i)}'^{-\varepsilon_i} S_i' S_{\lambda(i)}'^{\varepsilon_i}.$$

We introduce $S_i'' = S_i'^{-1}$ as new generators; then the relations (1) can be written as

$$S''_{i+1} \; S''^{\varepsilon_i}_{\lambda(i)} \; S''^{-1}_i \; S''^{-\varepsilon_i}_{\lambda(i)}$$

or

$$S''^{-1}_{i+1} \; S''^{\varepsilon_i}_{\lambda(i)} \; S''_i \; S''^{-\varepsilon_i}_{\lambda(i)} \, .$$

It follows from a comparison with the relations (2) in §3, Ch. III, that the group of the inverse knot is isomorphic to the group of the original knot.

We obtain a knot projection k' of the mirror image of a knot k by changing the undercrossings of the projection into overcrossings, and vice versa.

If we now rotate the projection plane of k' through 180° about a line g lying in it, then we obtain a projection k" of the mirror image knot, which we recognize as the mirror image of the original projection with respect to the line g.

We assign the generators S''_i ($i = 1, 2, \ldots, n$) to the projection k" so that S''_i and S_i are corresponding mirror image arcs. Then the relations among the S''_i read as follows:

(2) $$S''^{-1}_{i+1} \; S''^{-\varepsilon_i}_{\lambda(i)} \; S''_i \; S''^{\varepsilon_i}_{\lambda(i)} .$$

If we take the $S'_i = S''^{-1}_i$ as generators, then we obtain for the relations (2),

$$S'^{-1}_{i+1} \; S'^{\varepsilon_i}_{\lambda(i)} \; S'_i \; S'^{-\varepsilon_i}_{\lambda(i)} .$$

It follows from a comparison with the relations (2) in §3, Ch. III, that the group of the mirror image knot is isomorphic to the group of the knot.

Nevertheless, it can be established with the aid of the groups, for instance, that the torus knots are not amphicheiral (cf. §12, Ch. III).

§6. The matrix $(\ell_{ik}(x))$ and the group

We can now easily establish the group-theoretic significance of the matrices that were defined in chapter two ([3],[29]). Recall from §3, Ch. III, the quotient of a knot group W by its commutator subgroup K is infinite cyclic. We can thus choose as representatives of the residue classes of W modulo K the powers $S^{\ell}(\ell = 0, \pm 1, \ldots)$ of some generator, say $S = S_1$. We will produce the generators and defining relations of K by using the method for determining the generators and defining relations of subgroups given in ([27],[33]). We first introduce in place of the generators S_k of the knot group new generators E_k by means of

$$S_k = E_k S_1 \quad (k = 2, 3, \ldots, n);$$

clearly the E_i belong to the commutator group. Furthermore, we pick E_1 to be the unit element E. Relations (2) of §3, Ch. III, written in terms of the new generators have the form

$$R_i(E,S) = S_1^{-1} E_{i+1}^{-1} (E_{\lambda(i)} S_1)^{\varepsilon_i} E_i S_1 (E_{\lambda(i)} S_1)^{-\varepsilon_i}.$$

Using the algorithm cited above, the generators of the commutator group have the form

$$S_1^{\ell} E_k S_1^{-\ell} = E_{k\ell} \quad (k = 2, 3, \ldots, n; \ell = 0, \pm 1, \ldots)$$

(here, $E_{k0} = E_k$) and the relations are expressions of $S_1^\ell R_k(E_k, S_1) S_1^{-\ell}$ in terms of the $E_{k\ell}$. For $\varepsilon_i = +1$, one obtains

(1) $$R_{i\ell}(E) = E_{i+1,\ell-1}^{-1} E_{\lambda(i),\ell-1} E_{i\ell} E_{\lambda(i)\ell}^{-1}$$

and for $\varepsilon_i = -1$ one obtains

(2) $$R_{i\ell}(E) = E_{i+1,\ell-1}^{-1} E_{\lambda(i),\ell-2}^{-1} E_{i,\ell-2} E_{\lambda(i),\ell-1} \qquad \begin{aligned} &(i = 1,2,\ldots,n; \\ &\ell = 0, \pm 1, \ldots).\end{aligned}$$

The assignment $SKS^{-1} = K'$ gives an automorphism of the commutator group. We now make K commutative and introduce the operator x, by setting

$$SKS^{-1} = K^x,$$

and introduce formal exponents

$$f(x) = a_n x^n + a_{n+1} x^{n+1} + \ldots + a_{n+m} x^{n+m}$$

(a_i, m, n are integers, $m > 0$) by setting

$$K^{f(x)} = (K^{a_n})^{x^n} (K^{a_{n+1}})^{x^{n+1}} \ldots (K^{a_{n+m}})^{x^{n+m}}.$$

We thus obtain from the commutator group a commutative group $K(x)$ with operators. Then $E_{k\ell} = E_k^{x^\ell}$ and the relations (1) and (2) become

$$R_{i\ell}(E) \sim E_{i+1}^{-x^{\ell-1}} E_{\lambda(i)}^{x^{\ell-1} - x^\ell} E_i^{x^\ell} \qquad (\varepsilon_i = +1),$$

$$R_{i\ell}(E) \sim E_{i+1}^{-x^{\ell-1}} \; E_{\lambda(i)}^{-x^{\ell-2}+x^{\ell-1}} \; E_i^{x^{\ell-2}} \quad (\varepsilon_i = -1).$$

All of the relations with equal i can be omitted except one; we keep the one with $\ell = 1$ and $\varepsilon_i = +1$ and the one with $\ell = 2$ and $\varepsilon_i = -1$. This is true since in the operator group, all relations with equal i are consequences of the relation with $\ell = 1$ or with $\ell = 2$ respectively. The exponent matrix of the defining relations of $K(x)$,

$$(3) \quad \begin{cases} R_{i,x}(E) = E_{i+1}^{-1} \; E_{\lambda(i)}^{1-x} \; E_i^x & (\varepsilon_i = +1), \\ \\ R_{i,x}(E) = E_{i+1}^{-x} \; E_{\lambda(i)}^{x-1} \; E_i & (\varepsilon_i = -1). \end{cases}$$

is identical with the matrix that arises purely formally from $(\ell_{ik}(x))$, by striking out the first column. Thus we obtain: The L-equivalence classes of the matrix $(\ell_{ik}(x))$ defined in §14, Ch. II, characterize the abelian group $K(x)$ with the operator x.

Each of the relations R_{i_0} is according to §9, Ch. III, a consequence of the remaining R_i ($i \neq i_0$). One can also omit one of the relations (3) and strike out an arbitrary row in the exponent matrix of relations (3), without changing the elementary divisors. We thus can recognize the L-polynomial of a knot defined in §14, Ch. II, as a knot property.

At the same time, we see how one can form other matrices with the same L-equivalence classes: e.g., if one starts with a different set of defining relations for the knot group, one obtains different defining relations for the group $K(x)$. One thus obtains, for example, the matrices given by J. W. Alexander (cf. §14, Ch. II) by starting with the generators (5) from §3, Ch. III, and the defining relations (6) from §3, Ch. III.

§7. The knot group and the matrices $(c_{\alpha\beta}^{h})$

It follows from the decomposition of the knot group with respect to its commutator group that the collection of all elements contained in the cosets $KS^{\ell h}$ ($\ell = 0, \pm 1, \ldots$) form a normal subgroup W_h.

A complete system of representatives of the cosets $W_h F$ is given by

$$E, S_1, S_1^2, \ldots, S_1^{h-1}.$$

Consequently, using the method cited in §6, Ch. III, we obtain as the generators of W_h, $S_1^k E_i S_1^{-k} = E_{ik}$ ($i = 2, 3, \ldots, n; k = 0, 1, \ldots, h-1$) and $H = S_1^h$. One obtains a complete system of relations if we express

$$S_1^k R_i S_1^{-k} = 1 \quad (k = 0, 1, \ldots, h-1)$$

in terms of the $E_{i\ell}$ and S_1^h. When $\varepsilon_i = +1$ this yields

$$E_{i+1,k-1}^{-1} E_{\lambda(i),k-1} E_{i,k} E_{\lambda(i),k}^{-1} = 1 \quad (k = 1, 2, \ldots, h-1)$$

and for $k = 0$

$$S_1^{-h} E_{i+1,h-1}^{-1} E_{\lambda(i),h-1} S_1^h E_{i,0} E_{\lambda(i),0}^{-1} = 1.$$

Ch. III, §7   KNOT GROUP AND THE MATRICES $(c_{\alpha\beta}^h)$ 89

When $\varepsilon_i = -1$ this yields

$$E_{i+1,k-1}^{-1} \, E_{\lambda(i),k-2}^{-1} \, E_{i,k-2} \, E_{\lambda(i),k-1} = 1 \quad (k = 2,3,\ldots,h-1)$$

and for $k = 0,1$

$$S_1^{-h} \, E_{i+1,h-1}^{-1} \, E_{\lambda(i),h-2}^{-1} \, E_{i,h-2} \, E_{\lambda(i),h-1} \, S_1^h = 1,$$

$$E_{i+1,0}^{-1} \, S_1^{-h} \, E_{\lambda(i),h-1}^{-1} \, E_{i,h-1} \, S_1^h \, E_{\lambda(i),0} = 1.$$

In this, if one sets $H = S_1^h = 1$, and abelianizes the resulting group, then one obtains as exponent matrix the matrix given in (8) of §2, Ch. II, which arose from $(c_{\alpha\beta}^h)$ by striking out the columns $(1,k)$ ($k = 0,1,\ldots,h-1$).

The group K_h which is obtained from W_h by adjoining the relation $H = 1$ is (as a comparison of the relations with those of the commutator group shows) isomorphic ([29]) to the group which arises from the commutator group by the adjunction of the relations

$$E_{i\ell} = E_{i,\ell+h} \quad (\ell = 0, \pm 1, \ldots).$$

From this it follows that there is a corresponding connection between the groups $K(x)$ and $K_h(x)$ which arise from K respectively K_h, by abelianizing K respectively K_h, and introducing the operator

$$S_1 K S_1^{-1} = K^x, \text{ respectively } S_1 K_h S_1^{-1} = K_h^x.$$

The relations of $K_h(x)$ result from those of $K(x)$ by adjoining the relations

(1) $$E_i^{x^h-1} = 1.$$

We can derive from the relations of $K_h(x)$ the matrices whose elementary divisors are the torsion numbers by introducing as before

$$E_{i\ell} = E_i^{x^\ell} \quad (\ell = 0, 1, \ldots, h-1)$$

as generators. From (1), we then have

$$E_i^{x^k} = E_{i\ell}, \quad \text{where} \quad k \equiv \ell \pmod{h}.$$

Thus the relations among the $E_{i\ell}$ which follow from (1) are satisfied. In the remaining relations we replace $E_i^{x^k}$ by $E_{i\ell}$. Each relation R of $K_h(x)$ gives rise to the h distinct relations which are produced by the introduction of the $E_{i\ell}$ in the relations R^{x^i} ($i = 0, 1, \ldots, h-1$).

It follows from these considerations that the torsion numbers are determined by the equivalence class of the matrix $(\ell_{ik}(x))$.

§8. The edge path group of a knot

The group of a knot is isomorphic to the fundamental (or the edge path) group of the complement of the knot. Recall that the complement of the knot is the manifold which arises from Euclidean 3-space upon removal of the points of the knot. We first define the edge path group, and then show that this group has n generators S_i with the defining relations (2) of §3, Ch. III.

In the Euclidean 3-space that contains the knot, we choose a fixed point A in the complement of the knot and consider the directed closed polygonal paths emanating from this point which do not intersect the knot. We denote an edge path which passes through A by w. We understand $w_1 w_2$ to be the path which arises from w_1 by attaching w_2 to it. w^{-1} is understood to be the path which is directed oppositely to w.

Paths w and w' will be called <u>homotopic</u> if each can be carried into the other by applying operations of the following type a finite number of times.

$\Delta.\alpha$. Let PQ be a segment of the path with endpoints P and Q. Furthermore, let PR and RQ be two additional segments with endpoints P, R, and R,Q. Suppose that the triangle PRQ has no point in common with the knot. Then PQ is replaced by PR, RQ. The triangle PRQ may degenerate to a segment.

$\Delta'.\alpha$. is the operation inverse to $\Delta.\alpha$. In this deformation process, self-intersection of the path is permissible.

A group element of the edge path group of the complementary space of a knot with the base point A is defined to be a class [w] of homotopic paths, based at A. By the product of two group elements $[w_1]$ and $[w_2]$ we understand the element $[w_1 w_1]$.

It is clear that this defines a group. The unit element is the class of paths which can be retracted to A; the inverse of an element [w] is the class $[w^{-1}]$ of paths directed in the reverse sense.

The structure of this group does not depend on A.

By $\{w\}$ we will understand the class of loops which are freely homotopic to w (arbitrary initial point). Each class $\{w_0\}$ then corresponds to a class of elements $[w'] [w_0] [w'^{-1}]$ of an edge path group, which arise from a definite element $[w_0]$ by means of conjugation with an arbitrary edge path $[w']$. The paths which can be collapsed to a point are also called <u>null homotopic</u>.

§9. Structure of the edge path group

In order to obtain the generators and defining relations of the edge path group, we arbitrarily fix a regular projection direction. Each point P of the knot has a projection ray that passes through it. After a choice of a positive direction, each such ray is decomposed by P into an upper and a lower half ray. We keep the lower half ray.

The lower half rays generate a half-cylinder z which is bounded by the knot (Fig. 28); the double generators (double axes) of the cylinder correspond to the double points of the knot projection; the cylinder crosses itself there. To the arcs s_i of the knot, there correspond strips of surface z_i of the cylinder which are bounded by s_i and the consecutive double axes emanating from the knot ($i = 1, 2, \ldots, n$). A left and right side of each z_i is defined by the orientation of the knot. Assume that the base point A does not lie on z ([34]).

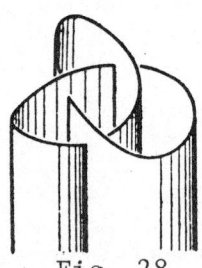

Fig. 28.

We first observe that two paths w_i and w_i' are homotopic if they pierce z only once, and both pierce the same z_i in the same direction (for example, from left to right. We denote the class $[w_i]$ by S_i.

Let w be an arbitrary path based at A. We can deform w so that it consecutively pierces the surface strips

$$z_{k_1}, z_{k_2}, \ldots, z_{k_r}.$$

That is, w decomposes into a product of elementary paths passing through A each of which pierces some surface strip z_i once. Therefore,

$$[w] = S_{k_1}^{\varepsilon_1} S_{k_2}^{\varepsilon_2} \ldots S_{k_r}^{\varepsilon_r},$$

where $\varepsilon_i = +1$ or $\varepsilon_i = -1$, according as z_{k_i} is pierced from left to right or from right to left. Since we can deform each path so that it pierces z only finitely often and does not cross any double axes, the S_i form a system of generators of the edge path group.

Naturally, formally distinct words may determine the same group element. This occurs if and only if the associated edge paths are homotopic, i.e., arise from each other by successive applications of $\Delta.\alpha$ and $\Delta'.\alpha$.

We can compose these deformations from those in which the surface of the triangle PRQ is met by at most one double axis and the boundary of the triangle pierces the half-cylinder z either exactly four times or exactly twice, depending on whether the triangle surface crosses a double axis ($\Delta.\alpha.1.$) or not ($\Delta.\alpha.2.$). The operations $\Delta.\alpha.2.$ cause the insertion or the deletion of the factor $S_i^{\varepsilon} S_i^{-\varepsilon}$. The operation

$\Delta.\alpha.1$. (Fig. 29) induces the application of a relation (2) from §3, Ch. III.

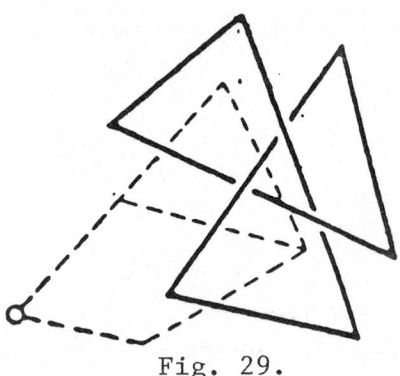

Fig. 29.

It follows that the relations (2) of §3, Ch. III, are the defining relations of the edge path group in the generators $S_i = [w_i]$.

We note further: Of the n relations $R_i(S)$, each is a consequence of the other $n-1$; for, a path which links a double axis once can be viewed instead as a path which links once each of the other $n-1$ double axes. The elements L_{n+1} defined in (2) of §4, Ch. III, can be represented by paths which correspond to a parallel knot consisting of one string. The homotopy class of an arbitrary parallel knot k_{qr} corresponds to a class of conjugate elements of the form

$$wL_{n+1}^{\pm q} S_1^r w^{-1}.$$

We can present the generators and defining relations of the path group in many ways. These may be obtained by cutting up the complementary space using surfaces other than the half cylinder z ([15]). For example, let β_w be the band defined in §4, Ch. I, which corresponds to the white regions, and which is bounded by a knot, and let β_s be the analogous band which corresponds to the black regions (including Γ_0). These bands can be so situated that they intersect in precisely n segments $d_i = U^i U_i$, each of which connects an overcrossing point with the corresponding undercrossing point. The knots and the segments d_i decompose the bands β_w and β_s into n+2 disks $f_0, f_1, \ldots, f_{n+1}$ which project onto the regions $\Gamma_0, \Gamma_1, \ldots, \Gamma_{n+1}$ respectively. We understand f to be the surface formed from the f_i (i = 1,2,...,n+1) (Fig. 30). For each path w there is a path w' that is homotopic to w and which pierces f only finitely often. We conclude from this that the classes

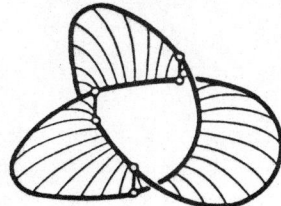
Fig. 30a.

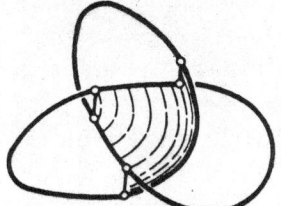
Fig. 30b.

$[w_i] = T_i$ (i = 1,2,...,n+1) of paths w_i which pierce f exactly once in f_i (i = 1,2,...,n+1) from under to over, form a system of generators for the edge path group; we may recognize that the defining relations in (6) from §3, Ch. III, correspond to those paths r_i that link the segments d_i once.

If w is a path that represents the class [w] = W and if W lies in the coset KS^v relative to the commutator group, then v is the intertwining number defined in §1, Ch. II, of w and the knot polygon. The subgroups W_h (see §7, Ch. III) therefore consist of the classes w of paths which link the knot kh times (k = 0,±1,...). If [w] lies in K and if [w] is distinct from the unit element, then the link consisting of $w = k_1$ and the knot $k = k_2$ has the intertwining number v equal to zero and linking number C_{12} not equal to zero.

§10. Covering spaces of the complementary space of the knot

There is in intimate connection between the edge path group and the unbranched covering spaces of a manifold which we shall describe in the case of the complement $\underline{A}$ of a knot.

Let z be the half cylinder described in §9, Ch. III, which is constructed by hanging parallel half rays along the knot k.

Euclidean 3-space is compactified by the adjunction of a point at infinity and is then cut open along this cylinder z. One obtains in this manner a cell Z whose boundary surface consists of two copies of the cylinder. The boundary surface is decomposed by k into two "pieces" z_λ and z_ρ and with the aid of the half rays corresponding to double points we decompose the boundary surface into the pieces $z_{\lambda i}$, $z_{\rho i}$ ($i = 1, 2, \ldots, n$). Now we take a finite or countably infinite number of copies of Z; we label the k-th copy $Z^{(k)}$, and we label the surface pieces of their boundaries by $z_{\lambda i}^{(k)}$, $z_{\rho i}^{(k)}$ ($k = 1, 2, \ldots$). A covering Y of the complementary space $\underline{A}$ is defined by specifying a successive pairing of the surface pieces $z_{\lambda i}^{(k)}$ and $z_{\rho i}^{(\ell)}$ with the same index i and the same or different indices k and ℓ. Furthermore, we do this so that exactly four such surface piece pairs are grouped about each line over a double axis; hence if r_i is a closed path of $\underline{A}$ that links a double axis exactly once, then any path $r_i^{(j)}$ of the covering space Y that lies over r_i is closed.

Ch. III, §10   COVERING SPACES   97

By traversing the surface pieces corresponding to a z_i in a definite direction (say, from left to right) one goes from an arbitrary cell $z^{(j)}$ to the same or a different cell $z^{(k_j)}$. In this, the k_j run through all the indices of the cells $z^{(j)}$ and the numbers corresponding to one another

$$\begin{pmatrix} 1, 2, \ldots \\ k_1, k_2, \ldots \end{pmatrix} = \pi_i$$

form a particular permutation. Exactly one permutation π_i corresponds to each surface piece z_i, and by virtue of the special conditions on our covering space, the π_i satisfy the relations of the fundamental group of the complement of the knot. To be precise, if the generators S_i of the knot group W correspond to the surface pieces z_i and the relation

$$R_i(S) = S_{i+1}^{-1} S_{\lambda(i)}^{\varepsilon_i} S_i S_{\lambda(i)}^{-\varepsilon_i}$$

corresponds to the path r_i about a double axis, then

$$\pi_{i+1}^{-1} \pi_{\lambda(i)}^{\varepsilon_i} \pi_i \pi_{\lambda(i)}^{-\varepsilon_i}$$

must be the identity permutation. This follows since the paths $r_i^{(j)}$ are closed paths over r_i. Therefore the π_i generate a group p that is the homomorphic image of the knot group, and conversely each group p

of permutations that is the homomorphic image of the knot group yields a covering space Y having the required property.

Each point $A^{(i)}$ and each path $w^{(i)}$ of the covering space lies above a well-defined point A and above a well-defined path w of the complement of the knot. Conversely, if w is a path starting at A, then there is a unique path $w^{(i)}$ over w which starts at the point $A^{(i)}$ over A. Define operations $\Delta.v.$ on closed paths in Y to be similar to the operations $\Delta.\alpha.$ in $\underline{\underline{A}}$. Thus we may define homotopy of paths in Y, and hence the path groups in Y can be defined.

If w is a loop based at A, that is null homotopic in $\underline{\underline{A}}$, then each lifting $w^{(i)}$ of w is also a loop and is null homotopic in Y. Therefore, Y is an unbranched covering of $\underline{\underline{A}}$.

From this it follows: If w_1 and w_2 are homotopic and if $w_1^{(i)}$ and $w_2^{(i)}$ are liftings of w_1 and w_2 both starting at $A^{(i)}$ then either $w_1^{(i)}$ and $w_2^{(i)}$ are both loops or neither is a loop. The collection of the elements [w] from $\mathcal{W}$, for which the liftings $w^{(i)}$ of w starting at $A^{(i)}$ are loops, form a subgroup $U_Y^{(i)}$ of $\mathcal{W}$. The $U_Y^{(i)}$ (i = 1,2,...) form a complete class of conjugate subgroups. The edge path group of Y is isomorphic to the subgroup $U_Y^{(i)}$.

Conversely, a particular unbranched covering can be constructed for each complete class of conjugate subgroups $U^{(i)}$ of $\mathcal{W}$.

Thus, we may think of the subgroups of the edge path group of a knot as the edge path groups of the covering spaces of the complement of the knot. The covering Y_h corresponding to the groups $\mathcal{W}_h$ is obtained by setting

$$\pi_1 = \pi_2 = \ldots = \pi_n = \pi$$

and

$$\pi = \begin{pmatrix} 1,2,\ldots,h \\ 2,3,\ldots,1 \end{pmatrix}.$$

The elementary divisors of the matrices $(c_{\alpha\beta}^h)$ are the torsion numbers ([5],[27]) of the manifold Y_h.

The boundary of the covering space Y consists of one or more curves $k^{(i)}$ ($i = 1,2,\ldots,r$) which cover the knot k. One obtains a new covering space Y^* which has empty boundary if one includes in Y those boundary curves. One obtains the edge path group of Y^* from the edge path group W_Y of the corresponding Y by forming a system of paths $w^{(i)}$ ($i = 1,2,\ldots,r$) each of which link exactly one of the $k^{(i)}$ once, expressing $[w^{(i)}]$ in terms of the generators of W_Y, and adjoining these power products as defining relations. For example, one obtains $H = S^h = 1$ as a new relation for the edge path group of the space Y_h^*.

The analogous situation holds for links. Each three-dimensional manifold can be represented as a covering space Y^* of a link ([4]).

§11. The group of a parallel knot

When investigating the group of particular knots, it is frequently easier to obtain the group properties by using generators and relations that are different from the generators (1) from §3, Ch. III, and the relations (2) from §3, Ch. III. As an example, we give a system of generators and defining relations for the group W_{qr} of the parallel knot

k_{qr}. These new generators and relations can be obtained quite easily from the generators and defining relations of the group W of the original knot k. From this presentation we will obtain information about the structure of the group W_{qr} with respect to its relation with the group $W([14])$.

Let D_i ($i = 1,2,\ldots,n$) be the double points of a projection of k and let the generators and defining relations associated according to (1) and (2) of §3, Ch. III, be S_i ($i = 1,2,\ldots,n$) and

$$R_i(S_k) = S_{i+1}^{-1} S_{\lambda(i)}^{\varepsilon_i} S_i S_{\lambda(i)}^{-\varepsilon_i} \quad (i = 1,2,\ldots,n).$$

Let k_{qr} be the parallel knot with q strings and twisting number $r > 0$ that is constructed from k. Position k_{qr} so that the projection of the attached cylindrical braid with twisting number r lies between the set with q^2 double points that corresponds to D_n and the set with q^2 double points that corresponds to D_1. Using the method of §4, Ch. III, one can read off $q^2 n$ generators corresponding to the D_i's and an additional $r(q-1)$ generators that correspond to the double points of the cylindrical braid, for a total of $q^2 n + r(q-1)$ generators, and also $q^2 n + r(q-1)$ relations for the group W_{qr} of k_{qr}.

We will replace those generators by new ones that have a simple geometrical significance for the parallel knot k_{qr}. If we imagine that in place of the knot k we have the corresponding torus with k_{qr} lying on it, then we can assign to the generators S_k of W the corresponding linkings T_k of the torus; considered as a path from the group W_{qr}, each of the T_k ($k = 1,2,\ldots,n$) links the q strings of k_{qr}. From

Ch. III, §11   THE GROUP OF A PARALLEL KNOT   101

this interpretation of the T_i one gets that the relations $R_i(T_k) = 1$ ($i = 1,2,\ldots,n$) hold, which arise from $R_i(S_k)$ upon replacing each S_k by T_k. We now assert that W_{qr} is generated by the T_k ($k = 1,2,\ldots,n$) and one additional generator Q which corresponds to a path along the core of the torus, and W_{qr} has as defining relations, $R_i(T_k)$ ($i = 1,2,\ldots,n$), together with one new defining relation $R(Q,T_k)$ which we define later (see formula (16) below).

We consider first the $q^2 n$ double points $D_{i,k}$ ($i = 1,2,\ldots,n$; $k = 1,2,\ldots,q^2$) that correspond to the double points D_i. If $U_{i,k}$ are the corresponding undercrossing points, then we can select the enumeration so that the subarc of k_{qr} going from $U_{i,k}$ to $U_{i,k+q}$ contains no overcrossing points. The collection of generators that correspond to these arcs can clearly be eliminated. Thus, the generators that remain consist of those generators $S_{i,k}$ ($k = 1,2,\ldots,q$) that correspond to subarcs which go from a $U_{i-1,k}$ to a $U_{i,k}$ ($i = 2,3,\ldots,n$) together with those generators $S_{1,k}, S_{n+1,k}$ that correspond to the arcs which go through the cylindrical braid portion and end and begin at $U_{1,k}$ and $U_{n,k}$ respectively. If, for brevity, we set

(1) $$\prod_{k=q}^{1} S_{i,k} = T_i \quad (i = 1,2,\ldots,n+1),$$

then the relations which now correspond to the $D_{i,k}$ read as follows:

(2) $$S_{i+1,k} = T_{\lambda(i)}^{\varepsilon_i} S_{i,k} T_{\lambda(i)}^{-\varepsilon_i} \quad (i = 1,2,\ldots,n; k = 1,2,\ldots,q).$$

In this, ε_i is the characteristic (see (1), §2, Ch. II) of the double point D_i.

We now consider (1) as the defining equations of the new generators T_i. From (1) and (2) follow the relations

$$(3) \qquad R_i(T) = T_{i+1}^{-1} \, T_{\lambda(i)}^{\varepsilon_i} \, T_i \, T_{\lambda(i)}^{-\varepsilon_i} \qquad (i = 1,2,\ldots,n).$$

As a consequence of the cylindrical braid relations (as is also easily observed from the geometric significance of the elements T_i) we have that

$$(4) \qquad T_{n+1} = T_1.$$

Equations (1) for $i = 1,2,\ldots,n$ can be derived from equation (1) with $i = n+1$ by means of conjugations with suitable elements and applications of (2) and (3), and can thus be omitted. Therefore, of the equations (1) only

$$(1^*) \qquad R_{n+1,0} = \prod_{k=q}^{1} S_{n+1,k} \cdot T_1^{-1}$$

remains. We now replace (2) by the equivalent relations

$$(2^*) \quad \begin{cases} S_{i+1,k} = L_{i+1}(T)\, S_{1,k} L_{i+1}^{-1}(T), \quad L_{i+1}(T) = T_{\lambda(i)}^{\varepsilon_i} \, T_{\lambda(i-1)}^{\varepsilon_{i-1}} \cdots T_{\lambda(1)}^{\varepsilon_1} \\ \qquad\qquad (i = 1,2,\ldots,n;\ k = 1,2,\ldots,q). \end{cases}$$

Since the $S_{i,k}$ ($i \neq 1, n+1$) do not occur in the cylindrical braid relations, we eliminate these generators and retain from the equations (2*) only

(2**) $\quad R_{n+1,k} = S_{n+1,k}^{-1} L_{n+1}(T) S_{1,k} L_{n+1}^{-1}(T) \quad (k = 1,2,\ldots,q).$

We now turn to the cylindrical braid portions. The overcrossing segments decompose the braid into $r + 1$ layers each of q segments corresponding to the r twistings (to the right). Let the generators which correspond to the ℓ-th layer be denoted by

$$A_{\ell,k} \quad (\ell = 1,2,\ldots,r+1; \ k = 1,2,\ldots,q).$$

Let

(5) $\quad A_{1,k} = S_{n+1,k}, \quad A_{r+1,k} = S_{1,k} \quad (k = 1,2,\ldots,q);$

one obtains the relations in the form

(6) $\quad A_{\ell,k} = A_{\ell-1,1}^{-1} A_{\ell-1,k+1} A_{\ell-1,1} \quad (\ell = 2,3,\ldots,r+1; \ k = 1,2,\ldots,q)$

between the $A_{\ell,k}$ by the usual method after using the identities $A_{\ell,q} = A_{\ell-1,1}$. In this, the second indices are taken modulo q. One obtains from (6) the equations

(6*) $\quad A_{j+1,k-j} = \left(\prod_{\ell=1}^{j} A_{\ell,1} \right)^{-1} A_{1,k} \prod_{\ell=1}^{j} A_{\ell,1}.$

It follows from this, that, for $k = j + 1$,

$$\prod_{\ell=1}^{j+1} A_{\ell,1} = A_{1,j+1} \prod_{\ell=1}^{j} A_{\ell,1}$$

and therefore by repeated application of this relation, we get

$$\prod_{\ell=1}^{r} A_{\ell,1} = \prod_{\ell=r}^{1} A_{1,\ell}.$$

For brevity we now set

$$L = \prod_{\ell=r}^{1} S_{n+1,\ell}.$$

Taking (5) into consideration, then for $j = r$ it follows from (6*) that

(6**) $$S_{1,k-r} = L^{-1} S_{n+1,k} L \quad (k = 1, 2, \ldots, q),$$

while all other $A_{i,k}$ and all other relations from (6*) can be eliminated.

With the aid of (6**) we now eliminate the $S_{1,k}$ from $R_{n+1,k}$. We denote the elements $S_{n+1,k}$ by S_k. For brevity, we set

(7) $$L_{n+1}(T) L^{-1} = Q^{-1} \quad \text{and} \quad L = \prod_{\ell=r}^{1} S_\ell.$$

In addition to equations (3) and (4), we obtain as the defining relations of W_{qr} the following:

(8) $$R_{n+1,0} = T_1^{-1} \prod_{\ell=q}^{1} S_\ell$$

and

(9) $$R_{n+1,k} = S_{k+r}^{-1} Q S_k Q^{-1} \quad (k = 1,2,\ldots,q).$$

It follows from (9) that

(9a) $$S_k = Q^q S_k Q^{-q}.$$

We will now show that S_1 can be expressed in terms of Q and L_{n+1}. To this end, we observe that by (9)

(10) $$S_{k+tr} = Q^t S_k Q^{-t}$$

and hence taking (7) into consideration

$$S_{r+tr} S_{r-1+tr} \cdots S_{1+tr} = Q^t L Q^{-t},$$

$$S_{tr} S_{tr-1} \cdots S_{tr-r+1} S_{(t-1)r} \cdots S_1 = \prod_{\ell=t-1}^{0} Q^\ell L Q^{-\ell} = Q^t (Q^{-1} L)^t = Q^t L_{n+1}^t.$$

Now, let

(11) $$a = \rho r = \kappa q + 1;$$

then, on the one hand,

$$\prod_{i=a}^{1} S_i = Q^\rho L_{n+1}^\rho$$

and, on the other hand,

$$\prod_{i=a}^{1} S_i = S_1 (\prod_{i=q}^{1} S_i)^\kappa = S_1 T_1^\kappa.$$

Therefore,

(12) $$S_1 = Q^\rho L_{n+1}^\rho T_1^{-\kappa}.$$

Henceforth we consider equation (7) to be the definition of the new generator Q and we therefore eliminate the generators S_i. Next, by (10) and (11), we can set

$$S_{k+1} = Q^\rho S_k Q^{-\rho},$$

and

$$S_\ell = Q^{\ell\rho} S_1 Q^{-\ell\rho} \quad (\ell = 2, 3, \ldots, q).$$

If we introduce these expressions into $R_{n+1,k}$, there arise relations that follow from (9a) with $k = 1$. There consequently result from (7), (8), (9) the defining relations

(13) $$R_1' = T_1^{-1} Q^q (Q^{-\rho} S_1)^q, \quad R_2' = Q^{r\rho-1} (Q^{-\rho} S_1)^r L_{n+1}^{-1}$$

and

(14) $$R_3' = S_1^{-1} Q^q S_1 Q^{-q}.$$

Ch. III, §11   THE GROUP OF A PARALLEL KNOT   107

We now use (12) to eliminate S_1 from (13) and (14). Then observe that L_{n+1} and T_1 commute (by (3) and (4)). We obtain from (13) that

$$Q^{q\rho} = L_{n+1}^{-\rho q} T_1^{\kappa q+1}, \qquad Q^{r\rho-1} = L_{n+1}^{-\rho r+1} T_1^{\kappa r},$$

or, by considering (11), that

$$Q^{q\rho} = (L_{n+1}^{-q} T_1^r)^\rho, \qquad Q^{q\kappa} = (L_{n+1}^{-q} T_1^r)^\kappa.$$

These two equations can be replaced by the one equation

(15) $$Q^q = L_{n+1}^{-q} T_1^r.$$

Since (15) holds, so does (14).

The calculations are similar for $r < 0$. We therefore obtain: The group of the parallel knot k_{qr} is generated by the elements T_k ($k = 1, 2, \ldots, n$) and Q and has for defining relations

(16) $$\begin{cases} R_i(T_k) = T_{i+1}^{-1} T_{\lambda(i)}^{\varepsilon_i} T_i T_{\lambda(i)}^{-\varepsilon_i}, & (i = 1, 2, \ldots, n), \\ R(Q, T_k) = Q^{-q} L_{n+1}^{-q} T_1^r. \end{cases}$$

The $R_i(T_k)$ arise from the relations (2) in §3, Ch. III, of the group of the original knot by replacing each S_k by T_k. L_{n+1} is defined by (2*). T_{n+1} is equal to T_1.

§12. The groups of torus knots

The groups of torus knots are obtained as a special case of the result in §11, Ch. III. As the carrier of the torus we take a double point free projected circle; then the group of the torus knot with q strings with twisting number r is generated by Q and T_1, and since $L_{n+1}(T) = 1$, the group has the one relation $R = Q^{-q}T_1^r$.

Concerning the structure of this group ([32]), it is easy to establish the following: $Q^q = T_1^r$ commutes with every element of the group, and if Z is any element that commutes with each element of the group, then $Z = Q^{kq}$ ($k = 0, \pm 1, \ldots$). Therefore, the subgroup generated by $Q^q = T_1^r$ is the center Z of W. Hence, the quotient group $F = W/Z$ is a group with generators Q, T_1 and with defining relations

$$R_1 = Q^q, \quad R_2 = T_1^r.$$

It is the free product of the subgroups generated by Q and T_1 respectively. It would be easy to miss the further possibility of representing F as a free product of finite cyclic subgroups. The only elements of finite order in F are namely

$$WQ^iW^{-1} \quad \text{and} \quad WT_1^kW^{-1}$$

(W arbitrary; $i = 1, 2, \ldots, q-1$; $k = 1, 2, \ldots, |r| - 1$).

We deduce from this that the relatively prime numbers q and $|r|$, as the maximal order of the elements of finite order, are characteristic numbers for F and hence for W. Furthermore, the automorphisms of F and hence those of W can be completely determined; the latter have the form

(1) $\qquad Q' = WQ^\varepsilon W^{-1}, \quad T_1' = WT_1^\varepsilon W^{-1} \qquad$ (W arbitrary; $\varepsilon = \pm 1$).

One can use these facts to completely classify torus knots. It is easy to see that two torus knots that are twisted uniformly with $q = a$, $|r| = b$, or with $q = b$, $|r| = a$ are isotopic. Consequently, two similarly twisted torus knots consisting of q respectively q' strings and with twisting number r respectively r' are isotopic if and only if $q = q'$ and $r = r'$ or $q = |r'|$ and $|r| = q'$.

In order to complete the classification of torus knots we will also show that knots which are twisted in different ways but have the same number pairs q, $|r|$ are not isotopic ([16],[32]). Since a knot corresponding to q,r that is twisted to the right is the mirror image of the knot with $q, -r$; thus, we assert that <u>no torus knot is amphicheiral</u>.

It follows from the definition of T_1 in (1) of §11, Ch. III, that in a torus knot the element defined in (2) from §4, Ch. III, is $L_{n+1} = T_1^r = Q^{-q}$. Thus the homotopy class that is associated with the element $WQ^q W^{-1} = Q^q$ can be represented by a path parallel to the torus knot.

We now observe that Q^q, represented by such a parallel curve p, induces an orientation of the knot, and hence Q^q cannot be deformed into an inversely directed parallel path.

The collection of all group elements that correspond to paths parallel to p and directed the same as p will be represented by

$$Q^q S^\ell \quad (\ell = 0, \pm 1, \ldots).$$

The collection of all group elements corresponding to paths parallel to p and directed oppositely to p are represented by

$$Q^{-q} S^\ell \quad (\ell = 0, \pm 1, \ldots),$$

where S is an arbitrary element of the group ([3],[9]).

Therefore, if p were deformable into an oppositely directed longitudinal path, then we would have to have in the group W

$$Q^q = W Q^{-q} S^k W^{-1},$$

and hence, since Q^q commutes with all elements, we would have $W S^k W^{-1} = Q^{2q}$. On the other hand, by (12) in §11, Ch. III, we can choose S equal to $Q^\rho T_1^{-\kappa}$; here $L_{n+1}(T)$ which was defined in (12) of §11, Ch. III, is equal to the identity. It follows that $(Q^\rho T_1^{-\kappa})^k$ represents the identity element in $F = W/Z$. But that contradicts the fact that F is the free product of the subgroups generated by Q and T_1 respectively.

The proof of our assertion is now as follows: If k can be deformed into its mirror image k' then there must be a homeomorphism T of Euclidean space which carries k into itself and reverses the orientation of Euclidean space. That is, T is the extension to Euclidean space corresponding to the deformation that carries k onto its mirror image k'. But such a mapping does not exist. For, if we let

$$A(Q) = Q', \quad A(T_1) = T_1'$$

be the automorphism corresponding to such a mapping T of k onto itself, then, by (1), we have

$$Q' = WQ^\varepsilon W^{-1}, \quad T_1' = WT_1^\varepsilon W^{-1} \quad (\varepsilon = \pm 1).$$

If $\varepsilon = +1$ then the directed knot k is mapped onto itself with preservation of orientation; for, Q^q, the curve parallel to the knot that is directed the same way, goes to $Q'^q = Q^q$, that is, into itself. The curves which link around k once in the positive sense, again go to such curves, since the residue class $KQ^\rho T_1^{-\kappa}$ relative to the commutator subgroup K goes to itself. Consequently, the corresponding mapping must preserve the orientation. But if $\varepsilon = -1$, then the directed curve k is mapped onto itself with reversal of the orientation; but also a curve which is positively linked is mapped to a negatively linked curve, and the orientation is therefore again preserved.

§13. The L-polynomials of parallel knots

Using the group-theoretic interpretation of the L-polynomial, one can obtain the L-polynomial by means of group calculations.

We give here the calculation of the L-polynomial for parallel knots which were considered in §11, Ch. III, and also in ([14]).

Instead of the T_i we introduce new generators $T_i = E_i T_i$ and replace formally $T_1^\ell E_i T_1^{-\ell}$ by $E_{i\ell}$ ($i = 1, 2, \ldots, n$; $\ell = 0, \pm 1, \ldots$). If the T_i are made commutative, then it follows from $R(Q, T_k)$ in (16) of §11, Ch. III, that

$$(1) \qquad Q^q = T_1^{q\omega + r} \quad \text{with} \quad \sum_{i=1}^{q} -\varepsilon_i = \omega.$$

Therefore, in W_{qr},

$$(2) \qquad Q^q = W(E_{i\ell}) T_1^{q\omega + r}.$$

It follows from (1) in §11, Ch. III, that T_1 belongs to the residue class $K_{qr} S^q$ of the commutator group K_{qr} in W_{qr}, where S represents an element that links once. It follows from (2) that Q is in the residue class $K_{qr} S^{q\omega + r}$. Since q and r are relatively prime, Q and T_1 can be replaced by a pair of primitive elements E_{n+1} and S_1 from the free group generated by Q and T_1 where also E_{n+1} belongs to K_{qr} and S_1 belongs to $K_{qr} S$. Then

$$(3) \qquad Q = Q(E_{n+1}, S_1) = W_1(E_{n+1}, \ell) S_1^{q\omega + r}$$

and

$$(4) \qquad T_1 = T_1(E_{n+1}, S_1) = W_2(E_{n+1}, \ell) S_1^q$$

Ch. III, §13 THE L-POLYNOMIALS OF PARALLEL KNOTS 113

with

$$E_{n+1,\ell} = S_1^{\ell} E_{n+1} S_1^{-\ell}. \quad (\ell = 0, \pm 1, \ldots).$$

In order to form $K_{qr}(x)$, we set

$$S_1 E_i S_1^{-1} = E_i^x \quad (i = 1, 2, \ldots, n+1),$$

make the E_i^x commutative, and determine the form of the new matrix.

Since

$$E_{i\ell} = T_1^{\ell} E_i T_1^{-\ell} = W_3(E_{n+1,\ell}) S^{\ell q} E_i S^{-\ell q} W_3^{-1}(E_{n+1,\ell})$$

$$(i = 1, 2, \ldots, n)$$

in $K_{qr}(x)$, we have that

$$E_{i\ell} = E_i^{x^{\ell q}}.$$

Therefore using the relations (16) from §11, Ch. III, one obtains the exponent matrix belonging to $K_{qr}(x)$ (defined in §6, Ch. III) by replacing x by x^q in the exponent matrix of the group $K(x)$ of the original knot, and then adjoining a new column corresponding to E_{n+1} and a new row corresponding to the new relation $R(Q, T_i)$. E_{n+1} really appears only in the new row.

Now it follows from (2), making use of (3) and (4), that

$$W(E_{i\ell}) T_1^{q\omega + r} Q^{-q} = W(E_{i\ell}) \cdot \{W_2(E_{n+1,\ell}) S_1^q\}^{q\omega + r} \{W_1(E_{n+1,\ell}) S_1^{q\omega + r}\}^{-q};$$

hence the exponent of E_{n+1} is independent of $W(E_{i\ell})$, since therein the index i runs only from 1 to n. It follows further from the definition of W_1 and W_2 in (3) and (4) that: The exponent of E_{n+1} is the L-polynomial of the torus knot that has q strings and twisting number $q\omega+r$. Thus, if $F(x)$ is the polynomial of k and $P_{q\omega+r,q}(x)$ is the polynomial of this torus knot, then the polynomial corresponding to k_{qr} is

(5) $$F_{qr}(x) = F(x^q) \cdot P_{q\omega+r,q}(x).$$

The evaluation of $P_{q\omega+r,q}(x)$ yields a cyclotomic polynomial

$$P_{q\omega+r,q}(x) = \frac{\{x^{q(q\omega+r)} - 1\}(x - 1)}{(x^q - 1)\{x^{(q\omega+r)} - 1\}}.$$

By means of (5) one can recognize the numbers q and $q\omega+r$ of a parallel knot to be invariants, when $F(x)$ possesses a factor $Q(x) \neq 1$ which is not a cyclotomic polynomial. Furthermore, (5) yields an algorithm for the calculation of polynomials of cable knots. From these polynomials one succeeds in reading off for the similarly twisted cable knots of the s-th order the characteristic series q_i, r_i (i = 1,2,...,s) which by §8, Ch. I, characterize the construction of these knots, and consequently the polynomials completely classify these knots ([14]).

§14. Several special knot groups

The generator T_1 used in §12, Ch. III, for the group of the torus knot can, by §11, Ch. III, be represented by a path which links once the torus on which the knot lies. Corresponding generators can be introduced for a knot which lies on a surface of a higher genus p. The knot will be situated to meet each line of a parallel bundle in at most two points. For such a knot, there is always a system of $p+1$ generators of which p correspond to paths which link the surface. A knot, whose projection determines m finite black regions, can always be situated on a surface of the above-mentioned sort of genus m. Of the $m+1$ generators mentioned, there are m elements T_i which by (5) of §3, Ch. III, are assigned to the black regions, while the last generator corresponds to any path which goes exactly once through the cylinder z defined in §9, Ch. III.

Using these generators, one can, for example, easily determine the group W_2 for an alternating pretzel knot and the group W_2^* which arises from W_2 upon adjunction of the relation $H = S^2 = 1$ ([27]). If there lie a_i ($i = 1,2,3$) overcrossings on the three two-stringed braid parts of the pretzel knot, then W_2^* expressed with the appropriate generators U_i ($i = 1,2,3$) has the relations

$$(1) \qquad R_1 = U_1^{a_1} U_2^{-a_2}, \quad R_2 = U_1^{a_1} U_3^{-a_3}, \quad R_3 = U_1 U_2 U_3.$$

In case all $a_i > 1$, then $U_1^{a_1} = U_2^{a_2} = U_3^{a_3}$ generate the center Z of W_2^*; the quotient group $F = W_2^*/Z$ can be represented by a discrete transformation group of the plane. The a_i arise as the characteristic numbers for F and consequently as invariants for the corresponding knots. Since alternating pretzel knots with the same number triple a_i (independent of the order) are isotopic, the complete classification of these knots with $a_i > 1$ is reduced to the question of which of these knots are amphicheiral.

The calculation of the L-polynomial for alternating pretzel knots is further facilitated if the generators are chosen as above ([29]). Here it is essential to distinguish two cases, according as $a_1 + a_2 + a_3$ is even (case 1) or odd (case 2). We restrict ourselves to knots with $a_3 = 1$. Then, in case 1, both a_1 and a_2 are odd. If we set

$$a_i = 2\alpha_i + 1 \quad (i = 1,2), \qquad \beta = (\alpha_1 + 1)(\alpha_2 + 1),$$

then

(2) $$L(x) = -\beta + (2\beta + 1)x - \beta x^2.$$

In case 2, let, say, a_1 be even and equal to $2\alpha_1$ and let a_2 be odd. The L-polynomial has the degree $g = a_2 + 1$, and

(3) $$L(x) = \alpha_1 + \alpha_1 x^g + (a_1 + 1) \sum_{i=1}^{g-1} (-x)^i.$$

In case 2, it follows that a_1 and a_2 turn out to be knot invariants; in case 1, on the other hand, we obtain only that β is a knot invariant which fact can also be determined from the second torsion numbers.

For the figure eight knot ($a_1 = 2$, $a_2 = a_3 = 1$), the commutator subgroup is a free group with two generators; consequently, the word problem can be solved for the group of the figure eight knot ([29]). The automorphism group of the figure eight knot group deserves special notice since the figure eight knot is amphicheiral ([16],[23]).

For a different example, assume that we have a knot where as a consequence of relations (2) from §3, Ch. III, all S_i except for two, S and S', can be successively eliminated. Hence in this case only one relation

$$R(S,S') = LSL^{-1}S'^{-1} \quad \text{with} \quad L = L(S,S')$$

remains. Instead of S' we introduce $K = S'S^{-1}$ and then obtain

(4) $$R(S,K) = LSL^{-1}S^{-1}K^{-1} \quad \text{with} \quad L = L(S,K)$$

as the new defining relation. We will now show: If one adjoins the relation $S^2 = 1$, then there arises from $\mathcal{W}$ a group $\mathcal{W}^*$ which is isomorphic to a dihedral group. It follows from (4) that

(5) $$KSKS = LSL^{-1}S^{-1} \cdot SLSL^{-1}S^{-1} \cdot S = LS^2L^{-1}.$$

Hence, if $S^2 = 1$, we can bring all elements either into the form K^i or into the form SK^i. If we now set $L = K^{\pm c}$ or $L = SK^{\pm c}$, then it follows that

(6) $$S^{2c+1} = 1.$$

In this, we can assume that $2c+1$ is greater than zero; if c is greater than zero then $2c+1$ is the second torsion number of this knot. The plaits with four strings defined in §6, Ch. I (and the pretzel knots with $a_3 = 1$, which can by §6, Ch. I, be deformed into plaits with four strings), yield examples of such knots ([30]).

If the knot k has two composite parts, k_1 and k_2, then the group of k is isomorphic with the free product of the groups W_i of k_i ($i = 1,2$) with amalgamated subgroups ([33]); the amalgamation subgroups are infinite cyclic groups which are each generated by one of the elements defined in (1) of §3, Ch. III, $S^{(1)}$ respectively $S^{(2)}$ from W_1 respectively W_2.

In order to investigate links more closely, it is expedient to consider the quotient group of higher commutator groups ([1]). Thus one can recognize, for instance, that links which are plaits with four strings consist of linked curves (Fig. 16) having intertwining number zero, hence then the linking number defined in §1, Ch. II, is also zero.

§15. A particular covering space

From each group of permutations p which is the homomorphic image of the group W of a knot one can construct by §10, Ch. III, a corresponding covering space Y, whose properties are characteristic for the knot. We consider one example of such a covering space for an arbitrary knot whose group can be generated by two elements K,S with the defining relation (4) from §14, Ch. III ([30]).

If the $2c+1$ defined by (5) of §14, Ch. III, is not 1, then let K respectively S correspond to the permutations

Ch. III, §15 A PARTICULAR COVERING SPACE 119

$$\pi^* = \begin{pmatrix} 1,2,\ldots,2c+1 \\ 2,3,\ldots,1 \end{pmatrix},$$

$$\pi_1 = \begin{pmatrix} 1 \\ 1 \end{pmatrix} \begin{pmatrix} 2 & ,2c+1 \\ 2c+1,2 \end{pmatrix} \begin{pmatrix} 3 & ,2c \\ 2c,3 \end{pmatrix} \cdots \begin{pmatrix} c+1,c+2 \\ c+2,c+1 \end{pmatrix}.$$

Clearly, π^* and π_1 generate a dihedral group P. Suppose that the permutations π_i ($i = 2,3,\ldots,n$) corresponding to the generators (1) of §3, Ch. III, have been calculated and that the covering Y has been constructed according to the rule in §10, Ch. III. We assert: <u>In Y there lie $c + 1$ curves $k^{(i)}$ over the knot k</u>. Let z_1 be the piece of surface of the cylinder in the knot complement <u>A</u> that corresponds to S_1. If w is a path that pierces z_1 once and if A is a point on w, then there are $2c+1$ points $A^{(1)},\ldots,A^{(2c+1)}$ in the covering space over A, and $c+1$ different closed curves over w. That is, if we go out from $A^{(1)}$ along the path $w^{(1)}$ that lies over w, then we return to $A^{(1)}$ after a one-fold covering of w. If we go out from another point $A^{(i)}$, then we return to the initial point after a two-fold covering of w. In the preceding, the curves which one describes going out from $A^{(2)}$ and $A^{(2c+1)}$, from $A^{(3)}$ and $A^{(2c)}$, etc., are the same. We shall call them $w^{(2)},\ldots,w^{(c+1)}$.

Now $\pi_2,\ldots,\pi_n$ arise from π_1 by conjugation with π^*, therefore they are constructed analogously, and hence one can construct exactly $c + 1$ such covering curves over each path w that pierces any cylinder piece z_i. But each of the transpositions of a π_i arises under this conjugation from a specific transposition of π_1. Consequently if we shove w along the entire knot so that it returns to its initial position after describing a knotted torus with k as core, then the $c+1$ accompanying liftings also describe $c+1$ disjoint tori which close when w again reaches its initial position. We denote by $k^{(i)}$ the curve over k that links the path $w^{(i)}$ once.

By §10, Ch. III, the edge path group F of Y with the base point $A^{(1)}$ is isomorphic to a subgroup U_1 of W. U_1 consists of those elements of W which correspond to permutations from p which send the symbol 1 to itself. The identity element and the K^ℓ ($\ell = 1,2,\ldots,2c$) are a complete system of representatives of W/U_1. The procedure given in §6, Ch. III, for determining subgroups ([27],[33]) gives as generators of U_1 and hence of F

$$(1) \qquad U_i = K^i S K^{-m_i}, \quad U_{2c+1} = K^{2c+1} \qquad (i = 0,1,\ldots,2c)$$

and gives as defining relations

$$(2) \qquad K^i R(S_1 K) K^{-i} = R_i(U) \qquad (i = 0,1,\ldots,2c).$$

In this, $m_i = 2c + 1 - i$.

<u>F is the group of a link consisting of $c+1$ curves.</u> In order to show this, we introduce new generators in place of (1) and indeed generators that correspond to curves each of which link once, exactly one of the liftings of the knot.

The generator S already has this property. We assert that instead of the remaining generators we can introduce as generators a finite number of elements

(3) $$Q_i S^2 Q_i^{-1} \quad (i = 1, 2, \ldots, r).$$

All elements of the form (3) belong in any case to F since the corresponding permutations are all equal to the identity which is due to the fact that $\pi_1^2 = 1$.

But by (5) of §14, Ch. III, since $S^2 = 1$ it follows that in W, $K^{2c+1} = 1$, $KSKS = 1$, and hence that

$$K^{i+1} S K^{i+1} K^{-i} S K^{-i} = 1 \quad (i = 1, \ldots, c-1).$$

These expressions can therefore be written as products of conjugates of S^2 and of conjugates of the relations (2). The latter are equal to 1 in the knot group, and if we omit them, then we obtain

$$K^{i+1} S K^{i+1} K^{-i} S K^{-i}$$

and K^{2c+1} expressed in terms of finitely many $Q_1 S^2 Q_1^{-1}$.

These equations (as do all relations in W) also hold in F, and if we introduce on the one hand these $Q_1 S^2 Q_1^{-1}$ and on the other hand

$$K^{-k} S^2 K^k \quad (k = 1, 2, \ldots, c)$$

as further generators, then we can determine successively from KSKS, S and $K^{-1}S^2K$ first KSK and then $K^{-1}SK^{-1}$, and from $K^2SK^2K^{-1}SK^{-1}$ and $K^{-2}S^2K^2$ first K^2SK^2 and then $K^{-2}SK^{-2}$, etc. Clearly, we can represent all generators (1) in terms of the new generators $Q_i S^2 Q_i^{-1}$, $K^{-k}S^2K^k$, S and $[K^{2c+1}]$; finally, $[K^{2c+1}]$ is eliminated.

But now we can give certain relations between the new generators. If Q_1 belongs to the coset modulo F represented by K^{q_1}, then

$$Q_1 S^2 Q_1^{-1} = Q_1 K^{-q_1} K^{q_1} S^2 K^{-q_1} K^{q_1} Q_1^{-1} = Q_1^* K^{q_1} S^2 K^{-q_1} Q_1^{*-1},$$

where Q_1^* belongs to F. Now if

$$Q_2 S^2 Q_2^{-1}$$

is an element where Q_2 belongs to the same coset as K^{q_1}, then one sees that for suitable Q_2^* from F,

$$Q_2 S^2 Q_2^{-1} = Q_2^* Q_1^{*-1} Q_1 S^2 Q_1^{-1} Q_1^* Q_2^{*-1}.$$

But one can furthermore see that

$$Q_1 S^2 Q_1^{-1} = M Q_2 S^2 Q_2^{-1} M^{-1}$$

holds in F, where M belongs to F. If Q_1 belongs to the coset FK^q and Q_2 belongs to the coset FK^{-q}, then

$$K^q S^2 K^{-q} = K^q S K^q \cdot K^{-q} S^2 K^q \cdot K^{-q} S^{-1} K^{-q}.$$

Hence, if we now adjoin the $c+1$ curves over the knot k to the covering space, then we must certainly set $S = 1$ and $K^i S^2 K^{-i} = 1$ ($i = 1, 2, \ldots, c$). In fact, it then follows from the statements just made that the resulting group becomes the identity.

It can be conjectured that all closed curves of Y can be deformed so that they run around in the interior of a particular cell. Specifically, the curves $k^{(i)}$ over k and all curves which correspond to the generators of F can be so deformed. Consequently the group F is in fact the group of a link. If we eliminate all the curves $k^{(i)}$ with the exception of $k^{(1)}$, then all generators with the exception of S equal 1. One can therefore conjecture that $k^{(1)}$ is unknotted.

Examples of knots with generators S, K, and one defining relation (4) of §14, Ch. III, are the plaits with four strings given in §6, Ch. I. A closer investigation is fruitful for the alternating plaits with four strings. Here, it turns out that the groups of the curves $k^{(i)}$ are free groups with one generator ([30]); the linking numbers[1] $v_{i\ell} = v_{\ell i}$ of each pair of oriented $k^{(i)}$, $k^{(\ell)}$ are equal to ± 2 or 0; v_{ii} is always equal to ± 2.

Other invariants of these knots can be obtained from the matrix $(v_{i\ell})$. Thus, for example, the sequence of the sums

$$\sum_{\ell=1}^{c+1} |v_{i\ell}| = 2v_i,$$

---

[1] C. Bankwitz, unpublished.

which states that $k^{(i)}$ is linked with v_i curves. The linking relations can be expressed by constructing a graph with $c+1$ points $P^{(i)}$ in which each pair $P^{(i)}$, $P^{(\ell)}$ is connected by a segment if and only if $v_{i\ell} \neq 0$.

Using the invariants v_i one can show that a_1 and a_2 are characteristic numbers[1] for the alternating pretzel knots where $a_1 = 2\alpha_1 + 1$, $a_2 = 2\alpha_2 + 1$, $a_3 = 1$ are the overcrossings on the two-stringed braid parts. From this it follows that the knots 7_4 and 9_2 of the table are not isotopic. Also the knots 8_{14} and 9_8 can be deformed into plaits with four strings (Fig. 15) and proved to be non-isotopic with the aid of the v_i. The v_i have the following values for:

7_4: 7, 5, 5, 4, 4, 3, 2, 2

9_2: 7, 6, 5, 4, 4, 3, 2, 1

8_{14}: 15, 12, 10, 10, 9, 9, 8, 8, 8, 7, 7, 7, 6, 5, 4, 3

9_8: 15, 12, 10, 10, 9, 9, 9, 8, 8, 7, 7, 6, 6, 5, 5, 2

Furthermore, the knots 9_{28} and 9_{29} can also be recognized to be non-isotopic by constructing a suitable three-sheeted covering space. The signs of the $v_{i\ell}$ are also characteristic when one uses the orientation of k to induce an orientation on the lifted curves $k^{(i)}$. From this there results a method for proving that the mirror images of alternating plaits with four strings (hence in particular, also of

---

[1] C. Bankwitz, Über die Torsionszahlen der alternierenden Knoten. Math. Ann. 103 (1930) 145-162.

alternating torus knots) are non-isotopic. For alternating torus knots the curves $k^{(i)}$ form a torus link.

TABLE OF KNOTS

The table of the following knot projections, up to nine overcrossings, was taken from the work of ALEXANDER and BRIGGS [5]. The curves 8_4 and 9_7 for which the number of overcrossings did not agree have been improved.

TABLE OF KNOTS

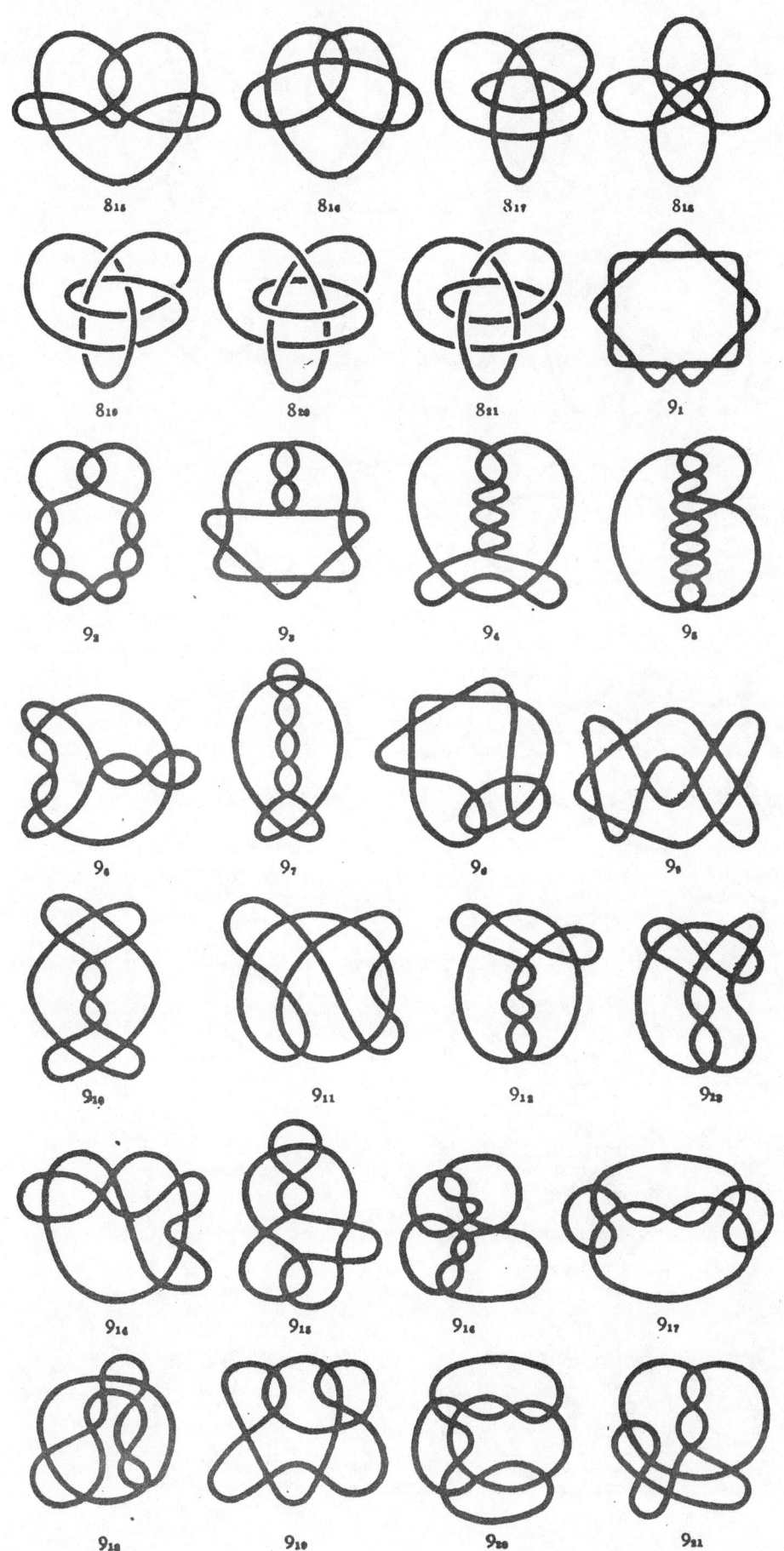

128 TABLE OF KNOTS

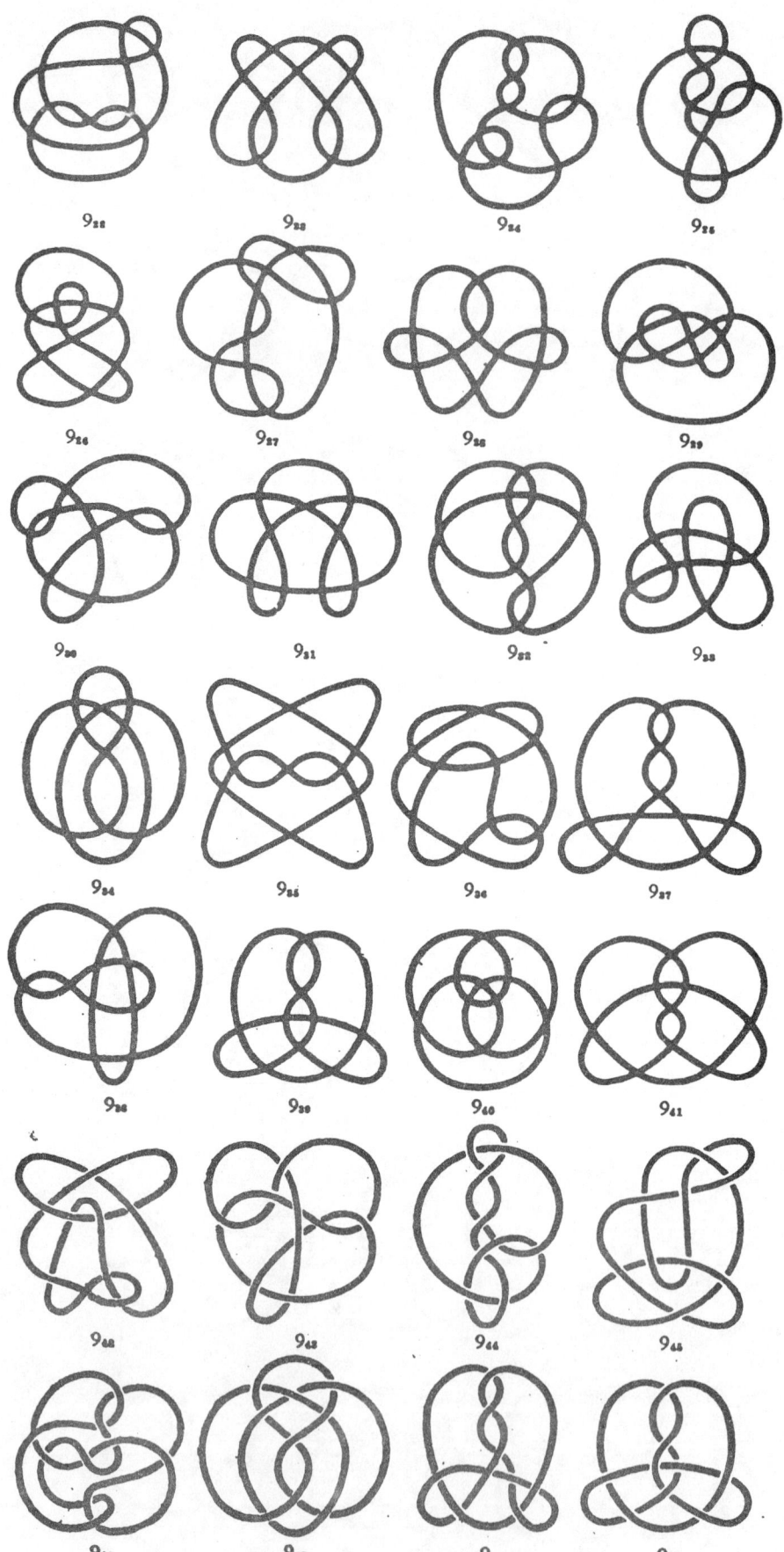

LITERATURE

[1]   ADELSBERGER, H.

Über unendliche diskrete Gruppen. J. Reine Angew. Math. 163 (1930) 103-124.

[2]   ALEXANDER, J. W.

A lemma on systems of knotted curves. Proc. Nat. Acad. Sci. U.S.A. 9 (1923) 93-95.

[3]   ALEXANDER, J. W.

Topological invariants of knots and links. Trans. Amer. Math. Soc. 30 (1928) 275-306.

[4]   ALEXANDER, J. W.

Note on Riemann spaces. Bull. Amer. Math. Soc. 26 (1920) 370-372.

[5]   ALEXANDER, J. W. and G. B. BRIGGS

On types of knotted curves. Ann. of Math. 28 (1926/1927) 562-586.

[6]   ANTOINE, L.

Sur l'homéomorphie de deux figures et de leurs voisinages. J. Math. Pures Appl. 4 (1921) 221-325.

LITERATURE

[7] ARTIN, E.

Theorie der Zöpfe. Abh. Math. Sem. Hamburg 4 (1925) 47-72.

[8] BANKWITZ, C.

Über die Torsionszahlen der zyklischen Überlagerungsräume des Knotenaussenraumes. Ann. of Math. 31 (1930) 131-133.

[9] BANKWITZ, C.

Über die Fundmentalgruppe des inversen Knotens und des gerichteten Knotens. Ann. of Math. 31 (1930) 129-130.

[10] BANKWITZ, C.

Über die Torsionszahlen der alternierenden Knoten. Math. Ann. 103 (1930) 145-162.

[11] BRAUNER, K.

Zur Geometrie der Funktionen zweier komplexen Veränderlichen. II. Das Verhalten der Funktionen in der Umgebung ihrer Verzweigungsstellen. Abh. Math. Sem. Hamburg 6 (1928) 1-55.

[12] BRUNN, H.

Topologische Betrachtungen. Zeitschrift für Mathematik und Physik 37 (1892) 106-116.

[13] BURAU, W.

Über Zopfinvarianten. Abh. Math. Sem. Hamburg 9 (1932) 117-124.

[14] BURAU, W.

Kennzeichnung der Schlauchknoten. Abh. Math. Sem. Hamburg 9 (1932) 125-133.

[15] DEHN, M.

Über die Topologie des dreidimensionalen Raumes. Math. Ann. 69 (1910) 137-168.

[16] DEHN, M.

Die beiden Kleeblattschlingen. Math. Ann. 75 (1914) 402-413.

[17] DEHN, M. and P. HEEGAARD

Analysis Situs. Enzyklopädie der Mathematischen Wissenschaften III A B Vol. 3 (1907) 153-220.

[18] FRANKL, E. and L. PONTRJAGIN

Ein Knotensatz mit Anwendung auf die Dimensionstheorie. Math. Ann. 102 (1930) 785-789.

[19] GOERITZ, L.

Knoten und quadratische Formen. Math. Z. 36 (1933) 647-654.

[20] HOPF, H.

Über die algebraische Anzahl von Fixpunkten. Math. Z. 29 (1929) 493-524.

[21] KÄHLER, E.

Über die Verzweigung einer algebraischen Funktion zweier Veränderlichen in der Umgebung einer singulären Stelle. Math. Z. 30 (1929) 188-204.

[22] KNESER, H.

Geschlossene Flächen in dreidimensionalen Mannigfaltigkeiten. Jber. Deutsch. Math.-Verein. 38 (1928) 248-260.

[23] MAGNUS, W.

Untersuchungen über einige unendliche diskontinuierliche Gruppen. Math. Ann. 105 (1931) 52-74.

[24] MINKOWSKI, H.

Über die Bedingungen, unter welchen zwei quadratische Formen mit rationalen Koeffizienten ineinander rational transformiert werden können. Ges. Abh. Bd. 1, p. 219, Leipzig, Teubner, 1911. J. Reine Angew. Math. 106 (1890) 5-26.

[25] MINKOWSKI, H.

Zur Theorie der Einheiten in den algebraischen Zahlkörpern. Nachrichten Ges. Wiss. Göttingen, 1900, pp. 90-93.

[26] PANNWITZ, E.

Dissertation. Berlin, 1931.

LITERATURE

[27]   REIDEMEISTER, K.

Knoten und Gruppen. Abh. Math. Sem. Hamburg 5 (1926) 7-23.

[28]   REIDEMEISTER, K.

Elementare Begründung der Knotentheorie. Abh. Math. Sem. Hamburg 5 (1926) 24-32.

[29]   REIDEMEISTER, K.

Über Knotengruppen. Abh. Math. Sem. Hamburg 6 (1928) 56-64.

[30]   REIDEMEISTER, K.

Knoten und Verkettungen. Math. Z. 29 (1929) 713-729.

[31]   ROHRBACH, H.

Bemerkungen zu einem Determinantensatz von Minkowski. Jber. Deutsch. Math.-Verein. 40 (1931) 49-53.

[32]   SCHREIER, O.

Über die Gruppen $A^a B^b = 1$. Abh. Math. Sem. Hamburg 3 (1923) 167-169.

[33]   SCHREIER, O.

Die Untergruppen der freien Gruppen. Abh. Math. Sem. Hamburg 5 (1927) 161-183.

[34]   WIRTINGER, W.

Über die Verzweigungen bei Funktionen von zwei Veränderlichen.

Jber. Deutsch. Math.-Verein. 14 (1905) 517.

The following papers treat analogous problems in higher dimensions:

ARTIN, E.: Zur Isotopie zweidimensionaler Flachen im R_4. Abh. Math. Sem. Hamburg 4 (1926) 174-177.

KAMPEN, E. R. van: Zur Isotopie zweidimensionaler Flächen im R_4. Abh. Math. Sem. Hamburg 6 (1928) 216.

SOMMERVILLE, D. M. Y.: On links and knots in Euclidean space of n dimensions. Messenger of Math. 36 (1907) 139-144.

Concerning the older literature the reader is referred to the Enzyklopädie der Mathematischen Wissenschaften article cited in [17].

Editors' Note: Concerning the literature up to 1967, the reader is referred to R.H. Crowell and R.H. Fox, An Introduction to Knot Theory, Springer-Verlag, New York, 1977; there is also an extensive bibliography at the end of the lead article in Springer-Verlag, Lecture Notes in Mathematics, vol. 285, 1978.

INDEX

Alexander polynomial of a knot 63-68

Almost alternating circles 61

Almost alternating knots 57, 58

Almost alternating projection 57

Alternating around a region 58

Alternating knot 6, 34, 55

Alternating knot projection 5

Alternating plait with four strings 123

Alternating pretzel knots 116, 125

Alternating torus knots 11, 18

Alternating torus link 18

Amphicheiral knot 2, 51, 85, 109

Black and white regions 10

Braid, closed 17

Braid, cylindrical 16, 17

Braid, cylindrical (with twisting r) 17

Braid group Z_q 73

Braid, open 15

Braid, order of a 17

Braid (open) with q strings 15, 73

Braid with two strings 11

Braid, torus 18

Braids 15, 19

Braids, equivalent (open) 71

Branched covering space 96

Cable knot 24, 75, 114

Characteristic, ε, of a double point 28

Characteristic numbers for alternating pretzel knots 115, 124

Circle 2

Classes of equivalent braids 73

Classification of alternating knots 55

Closed braid 17

Cloverleaf knot 11

Complement of a knot $\underline{\underline{A}}$ 96

Composite part of a knot 12, 118

Core of a parallel knot 22

Covering space, branched 96

Covering space, unbranched 96, 98

Covering spaces 96, 118

Crossing of the k-th and (k+1)-st strings of a braid 16

Cyclotomic polynomial 114

Cylindrical braid 16, 17

Cylindrical braid (with twisting r) 17

Dehn's lemma 26

Deformation of a link 2

Deformation (of a polygon) 1

Deformation of a projected curve 6

Deformation of an open braid 71

INDEX

Determinant, irreducible 52

Determinant, irreducible components of a 53

Determinant of a knot 33

Dihedral group 117, 119

d_{ij} (the number of double points in which Γ_i and Γ_j are contiguous) 34

Double points 3

Edge path group (of a knot) 27, 90, 92

Elementary (matrix) transformations ($\Sigma.\xi.1$, $\Sigma.\xi.2$, $\Sigma.\xi.3$, $\Sigma.\xi.4$, $\Sigma.\xi.5$) 67

Equivalent (open) braids 71

Euler-Poincaré characteristic 25

Exponent matrix 87, 89, 113

Figure eight knot group 117

Fundamental group of the complement of a knot 90

Gleichsinnig verdrillt 16

Group, dihedral 117, 119

Group, edge path (of the complement of a knot) 90, 92

Group elements 73

Group, fundamental (of the complement of a knot) 90

Group of a knot, $\mathcal{W}$ 76

Group of a link 76, 120

Group of a parallel knot 99

Group of a torus knot 108

Groups, amalgamated 118

Half-cylinder 92

Homotopic, null 92

Homotopic paths 91

Hyperboloid of singular projection directions 3

Incidence number, η 31

Intertwining number of a link 26

Invariance of the knot group 79

Invariance of torsion numbers 35

Inverse knot 2, 83

Irreducible components of a determinant 53

Irreducible determinant 52

Isotopic links 2

Isotopic polygons 2

Isotopic systems 2

Knot 2

Knot, alternating 6

Knot, amphicheiral 2, 109

Knot, cable 24, 75

Knot, cloverleaf 11

Knot, figure eight 117

Knot, inverse 2, 83

Knot, oriented 2

Knot, parallel 23, 99, 111, 114

Knot, pretzel 11, 38

Knot, prime 12

Knot property 2

Knot, s-th degree cable 24

Knot, symmetric 2

INDEX

Knot, torus 18, 85, 108, 109

Knot, trefoil 11, 75

Knotted polygon 2

Knottedness, measure of 25

Knotting number 4, 26

Legendre symbol 48

L-equivalent matrices 67

Link 2

Link, order of a 2

Link, torus 18

Linking number 4, 26, 27, 123

L-polynomial of a knot 63 - 68, (table) 70

L-polynomial of a parallel knot 111

L-polynomial of a torus knot 114

L-polynomial of an alternating pretzel knot 116

Matrices, L-equivalent 67

Matrix $(c_{\alpha\beta}^2)$ 31

Matrix $(c_{\alpha\beta}^h)$ 27, 88

Matrix (a_{ik}) 31

Matrix $(\ell_{ik}(x))$ 63, 85

Matrix (b_{ik}) 31

Matrix (b_{ik}') 31

Matrix $(b_{\alpha\beta}^h)$ 31

Matrix $(a_{k\ell}')$ 33

Matrix (v_{ik}) 123

Measure of knottedness 25

Minkowski's units, C_p 46-51

Minkowski's units (table of) 51

Mirror image knot 6, 83

m(k) 25

Non-compact region, Γ_0, of a subdivision 10

Normal knot projection (for two regions) 13

Normal projection 13

Normalization of a projection 5

Normalized Alexander polynomial 63

Null homotopic 92

Number, incidence 31

Number, intertwining 26

Number, knotting 4, 26

Number, linking 4, 26, 27, 123

Number, second torsion 38, 39, 118

Number, third torsion 39

Number, torsion 30

Number, torsion (second) 38, 39, 118

Number, torsion (third) 39

Open braid 15

Operations (on a polygon in Euclidean 3-space): Δ, Δ' 1

Operations (on a projected polygon):

    ($\Delta.\pi.1,2$) 6

    ($\Omega.1, 2, 3, 4, 5$) 6, 7, 8, 12, 13, 14

    ($\Omega'.1, 2, 3$) 8

Operations (on a closed braid) 18

Operations $\Delta.\zeta.\pi$ 71

Operations ($\Omega.\zeta.2, \Omega.\zeta.3$) 72

Operations (on the edge path group): $\Delta.\alpha, \Delta'.\alpha$ 91

Operations (on a projected braid) 17

Operations (in covering spaces): $\Delta.\nu$ 98

Operations ($\Delta.\zeta, \Delta'.\zeta$) 71

Operations ($\Omega.1\alpha, \Omega.2\alpha, \Omega.1\beta, \Omega.2\beta$) 40, 42

Operations ($\Sigma.1, \Sigma.2, \Sigma.3$) 46, 47

Order of braid 17

Order of a link 2

Orientable surface 10

Orientation in the projection plane 28

Oriented knot 2

Overcrossing point 5

Parallel knot 25, 99, 111, 114

Parallel knot, group of a 99

Parallel knot, L-polynomial of a 111

Parallel projection 3

Plait with four strings (Viergeflecht) 18, 21, 118, 123

Point, overcrossing 5

Point, undercrossing 5

Pretzel knot 11, 38

Prime knot 12

Principal minors (of a determinant) 52

Product of two classes 73

Projection, alternating (knot) 5

Projection, normal 13

Projection, normal (for two regions, Γ_i and Γ_k) 13

Projection, parallel 3

Projection, regular 3

Projection, schema of a 4

Property, knot 2

Quadratic form (of a knot) 40

Quotient group 76

Regular projection 3

Regions (white and black) 10

Schema of a projection 4

Second commutator group of a group 77

Second torsion number 38, 39, 118

s_i, s_i^{-1} [braid with overcrossing (undercrossing) string] 72

Singular points of a projection 3

Singular projection directions 3, 4

s-th degree cable knot 24

Strings of a braid 16

Subdivision of the projection plane 10

Surface spanned by a knot 10, 25

Symmetric knot 2, 22

Table of torsion numbers 39

Theorem of Bankwitz 56

Third torsion number 39

Torsion numbers 30, 38, 99

Torsion numbers, h-th 30

Torsion numbers, invariance of 35

Torsion numbers, table of 39

Torus braid 18

Torus knot   18, 85, 108, 109, 114

Torus knot, group of a   108

Torus link   18

Transformation problem   71, 74

Transformation of a projection direction   9

Transformations ($\Delta.\xi.1, 2, 3, 4, 5$)   67

Trefoil knot   11, 75

Twisted uniformly (gleichsinnig verdrillt)   16, 18

Unbranched covering space   96, 98

Undercrossing point   5

Units of a quadratic form   47

Unlinked polygons   27

Viergeflecht   18

Word problem   71, 74